IAIN NICOLSON

ASTRONOMY

EXPLORING THE NIGHT SKY

KT-376-088

TREASURE PRESS

FOREWORD

This guide tells the story of astronomy, the oldest of the sciences and yet a field in which many of the most exciting developments of our age are taking place. Astronomers of the distant past who had only their own eyes with which to make observations would be baffled by the complex equipment in use today, ranging from the smallest optical telescopes to huge radio telescopes which, though weighing thousands of tons, can nevertheless be directed accurately to any part of the sky.

The universe revealed by these instruments extends over distances in space and time which defy the imagination, and contains, besides the more familiar stars and planets, a great variety of other objects, such as the mysterious 'quasars' recently discovered and described later on in this book. This is an exciting time in astronomy, with men taking the first steps in manned exploration of our solar system and sophisticated space probes already sending back information from the Moon and the planets Venus and Mars. However, the amateur observer can still make useful contributions to the science of astronomy and some of the ways in which he can do this are described. Astronomy makes a fascinating pastime whether you make important observations with your own equipment or simply learn to recognize the constellations.

Astronomy is certainly not a static subject – new ideas and discoveries are always turning up. Although astronomers have solved many of the mysteries of the universe, they are continually finding new and deeper problems.

I.K.M.N

First published in Great Britain in 1970 by The Hamlyn Publishing Group

This edition first published by
Treasure Press
Michelin House
81 Fulham Road
London SW3 6RB

Printed in Hong Kong

CONTENTS

4	Introduction
20	Astronomical instruments
32	The solar system
38	The Sun
48	The planets and their satellites
68	Minor bodies of the solar system
76	The stars
92	Our galaxy
100	Galaxies and the universe
110	The future
118	The constellations
144	The amateur astronomer
156	Books to read
157	Index

INTRODUCTION

By the very act of looking at the sky we become involved with astronomy. There can be few people who have not gazed at some time at the stars, and pondered just a little on mysteries of the universe and man's place in it. Astronomy tells a fascinating story, a tale made all the more enthralling by the ease with which anyone can participate in astronomical observation. The history of astronomy abounds in valuable contributions made by amateurs, and even today, when professional astronomers work with expensive and highly sophisticated equipment, there is still a place for the amateur observer with his small telescope. It is possible to derive tremendous satisfaction from astronomy without any equipment but the naked eye; a good knowledge of the sky is a thoroughly pleasing thing, and the sight of the stars and planets shining against the background of the Milky Way in a clear sky is still perhaps the most moving experience in the whole of astronomy.

The Earth on which we live is simply one of a family of planets revolving around the Sun, while the Sun itself is no more than one very ordinary star in the huge stellar system – the Milky Way – which is our galaxy. Likewise, our galaxy is merely one minute speck among the vast hordes of galaxies which stretch to the edge of the observable universe – some of them so far distant that their light takes thousands of millions of years to reach us. In fact the observable universe is a hundred thousand million million million (written as 10^{23}) kilometres across! Against the background of the universe, the Earth seems very insignificant indeed.

Yet if we look to the other end of the scale, man does not seem so small. The atom, the basic building brick of matter, is only a ten thousand millionth of a metre in diameter (that is, 10^{-10} metres). The explanation of gigantic astronomical phenomena depends upon what goes on inside the tiny atom. Thus astronomy links the largest and the smallest things we know.

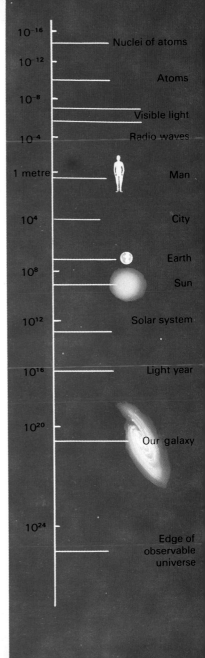

10^{-16}	Nuclei of atoms
10^{-12}	Atoms
10^{-8}	Visible light
10^{-4}	Radio waves
1 metre	Man
10^{4}	City
10^{8}	Earth / Sun
10^{12}	Solar system
10^{16}	Light year
10^{20}	Our galaxy
10^{24}	Edge of observable universe

The scale of the universe. Starting from the top, every division is 10,000 times greater than the preceding one.

Early watchers of the skies

Perhaps the most primitive of men saw patterns in the sky, but the recorded beginnings of astronomy certainly go back to around 3000 BC. Indeed, to the earliest tribes, the heavenly bodies were of the utmost practical importance. The Sun provided heat and light during the day, while the Moon provided light during the night. The regular behaviour of the Sun's rising and setting, the Moon's phases, and the positions of the Sun against the stars at different seasons gave man the means of regulating his life. Little wonder that the early watchers of the skies treated the Sun as a god and the Moon as goddess. The stars were lights set into a huge hemisphere beneath which lay the flat Earth, while eclipses, comets and shooting stars were all manifestations of the gods that controlled mankind's destiny.

Many of the ancient ideas seem strange nowadays. For example, some maintained that the flat Earth was supported on the backs of four huge elephants standing on the shell of an enormous tortoise, which was itself supported by a serpent floating in a boundless ocean.

The organized recording of astronomical events dates

Comets were thought by the ancients to be omens of the gods.

Aztec temple to the sun-god

Sundial

back to the earliest known great civilizations – in particular those that arose in China, India, South America and the Middle East. There are immense difficulties in analyzing the old records, often on stone tablets, but there is little doubt that the Chinese built sundials of an advanced type, and constructed – almost 5,000 years ago – a calendar of 365 days. They made careful records and discovered certain relationships which enabled them to predict some eclipses. The story is told of the luckless court astronomers, Hsi and Ho, who failed to predict an eclipse of the Sun, and were executed for endangering the safety of the world. Five 'wandering stars' or planets were certainly known in 2500 BC and some of these must have been discovered long before.

When an eclipse occurred, the Chinese thought that the Sun was being swallowed by a huge dragon. The whole population joined in making as much noise as possible to scare it away. They always succeeded

The Babylonians and Egyptians

Similiar heights of astronomical achievement were attained independently by other civilizations, but this knowledge tended to die out with their decline. Only in the case of the Babylonians and the Egyptians was information and astronomical skill passed on to others. The Babylonians made many accurate measurements of stars and planets; these were made not for the sake of 'science' in the modern sense, but for the purpose of divining the will of the gods, their gods being the Sun, the Moon and the planet Venus. They were notable mathematicians, developing a form of algebra and using the sexagesimal number system (based on 60, just as our decimal system is based on 10).

The Egyptian civilization grew up along the banks of the Nile. The fluctuations of this river were a matter of life and death to the early inhabitants. The prediction of the annual flood was of the utmost importance, and it soon became apparent that the lunar calendar then in use was not accurate enough. The Egyptians noticed that the coming of the flood coincided with the time when the star Sirius first became visible in the dawn sky and they drew up a very accurate

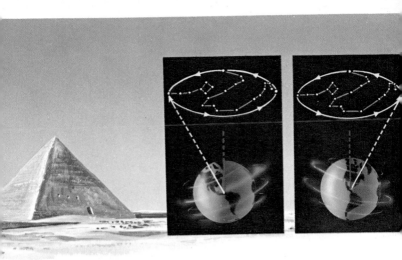

Like a spinning top, the Earth's axis precesses or wobbles. The Great Pyramid of Cheops was built with one passage pointing towards what was then the pole star, Thuban. Today the pole star is Polaris.

calendar based on the movements of this star. Their year had $365\frac{1}{4}$ days, though they employed a civil calendar of 365 days, consisting of twelve thirty-day months and five days' holiday.

An interesting point is that the Great Pyramid of Cheops is aligned with one main passage directed towards the north pole of the sky as it was at that time. The Earth's axis *precesses*, or wobbles, rather after the fashion of a spinning top, in such a way that the celestial north pole (that is, the point on the sky directly overhead at the Earth's north pole) describes a circle in the sky in a period of 26,000 years. The star which happens to lie very close to the celestial pole is known as the 'pole star'. At the present time it is the star Polaris. The Great Pyramid is aligned towards the star Thuban in the constellation Draco, which was indeed the pole star some 4,500 years ago.

However, the Egyptians treated their astronomical observations purely as a practical convenience, and made little or no advance in astronomical theory. Truly revolutionary ideas in astronomy were to await the coming of the Greeks.

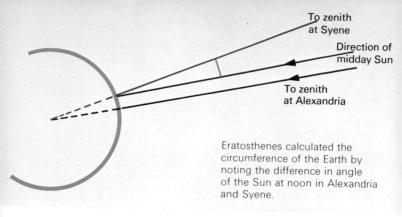

To zenith
at Syene

Direction of
midday Sun

To zenith
at Alexandria

Eratosthenes calculated the
circumference of the Earth by
noting the difference in angle
of the Sun at noon in Alexandria
and Syene.

The Greek astronomers

The foundations of science as we know it were laid in
Greece from 700 BC to AD 200. The Greeks were rational
thinkers and experimenters. Their development of geometry
was truly remarkable, and their contribution to astronomy
considerable.

Their first great astronomer was Thales of Miletus (born
640 BC), who believed the universe to be spherical; in the
Greek view, the circle and the sphere were perfect forms.
The very influential Aristotle (384–322 BC) put forward the
idea that the Earth was not flat, basing his views on two
important observations. Firstly, he knew that the positions
of stars in the sky changed as one went north or south, and
secondly, he had noticed that the shadow of the Earth on
the Moon during an eclipse was curved. Both these observations could be explained if the Earth were a sphere.

Soon afterwards Eratosthenes measured the circumference of the Earth and obtained a value of about 40,250 km –
very close to today's accepted value. His method was
beautifully simple: he measured the difference in angle
between the positions of the Sun at noon in the towns of
Alexandria and Syene. Having measured the actual distance
between the towns, it was then an easy matter to calculate the
distance around the globe (an angle of 360°).

Hipparchus (190–120 BC) was perhaps the greatest Greek
astronomer. Among his many achievements were the invention of trigonometry, his catalogue of the positions of stars,

and the discovery of the precession of the Earth's axis, described on the previous page. Ptolemy based his great astronomical book the *Almagest*, written in the second century AD, almost entirely on the theories of Hipparchus.

Although Aristarchus had suggested that the Earth might travel around the Sun, the final Greek view of the universe was crystallized by Ptolemy in the *Almagest*. The Earth lay at the centre of the universe, and around it revolved in circular orbits, in order of distance, the Moon, the planets Mercury and Venus, the Sun, the planets Mars, Jupiter, and Saturn, and finally the outer sphere of stars. The observed movements of the planets were not as regular as they should have been if they moved in circles, and Ptolemy introduced additional motions in the form of epicycles and deferents to overcome this flaw in his system.

The epicycle was an additional circle, smaller than the main circular orbit. The centre of the epicycle, the deferent, moved round the principal circle (with the Earth at the centre) while the planet moved round the epicycle and was carried along with it. The final form was thus rather complicated.

The Ptolemaic universe

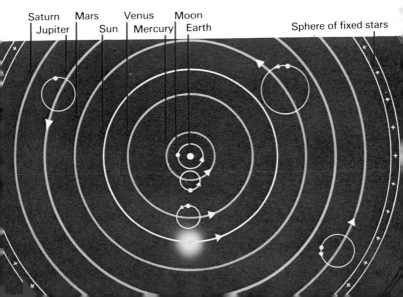

| Saturn | Mars | Venus | Moon | |
| Jupiter | | Sun | Mercury | Earth | Sphere of fixed stars |

The Dark Ages

After Ptolemy's time the great Greek civilization faded, and few further astronomical advances came out of Greece. Civilization slipped into the Dark Ages, science came virtually to a standstill, and the cosmology of Aristotle and Ptolemy remained largely unchallenged for over a thousand years. Unfortunately, the idea of a spherical Earth fell by the wayside during this period.

However, several centuries after Ptolemy's death the Arabs began to take a serious interest in astronomical observation. Many of the star names in use at present are of Arabian origin – for example, Aldebaran, 'the eye of the bull', a bright red star in the constellation of Taurus, the Bull. The Arabs measured the positions of stars with considerable accuracy. One fundamental step forward was their system of measuring stellar brightnesses, very similar to our present magnitude system (see page 78). However, they were more interested in astrology, and their accurate measurements were directed towards predicting the future rather than towards gaining pure astronomical knowledge.

The Renaissance

Suddenly, at least by comparison with the past long centuries of inactivity, a fresh interest in astronomy awoke in Europe. It was the fifteenth century, and the beginning of that great upsurge in human awareness – the Renaissance.

The year 1473 saw the birth of a man who was to shake the whole foundation of man's views of the universe – Nicolaus Koppernigk, or Copernicus. He made a study of the motions of the planets and rapidly came to the conclusion that the Ptolemaic system of circles and epicycles could not possibly account for the observed movements if the Earth were at the centre of the universe. Finally he decided that the Sun must be at the centre of the universe and that all the planets, Earth included, revolved around it. Such a concept seems obvious today, but at that time it was no less than shattering. In particular it ran contrary to the Church's belief that the Earth was the most important body, and anticipating the most severe consequences, he did not publish his theory until 1543, the year of his death, in the now famous book *De Revolutionibus Orbium Coelestium*.

(*Opposite left*) An Arabian astrolabe, used for observing the altitudes of the stars. (*Opposite right*) An armillary sphere. These devices were used in various forms by early astronomers. The rings represent the principal circles of the celestial sphere; the instrument was used for determining the positions of stars. (*Below*) Tycho Brahe's observatory. He made observations with precision never before achieved, providing the observational basis for Kepler's theories.

Tycho Brahe

Not all Copernicus' successors accepted his view. One of the most famous dissenters was the great Danish astronomer Tycho Brahe, who built and equipped a large observatory on the island of Hven. Here Tycho constructed maps of the stars and the motions of the planets which surpassed all previous efforts in their detail and accuracy. By all accounts, Tycho was an exceedingly unpleasant character – he is said to have lost part of his nose in a duel and to have replaced the missing part with a mixture of gold and wax – but his skill as an astronomer was astounding, especially as he had no telescopes to aid him. Tycho was utterly opposed to the *heliocentric* ('Sun-centred') theory of Copernicus, yet ironically enough his own observation enabled his pupil Johannes Kepler to prove conclusively that the Earth did indeed go round the Sun.

(*Above*) Tycho Brahe's quadrant, used for determining, in degrees, the altitudes of astronomical bodies.
(*Below*) Two of Galileo's telescopes as they are displayed today in Florence.

14

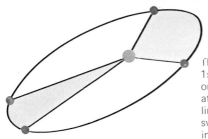

This diagram illustrates Kepler's 1st and 2nd laws. The planets' orbits are elliptical, with the Sun at one focus of the ellipse. The line joining a planet to the Sun sweeps out equal areas of space in equal times.

Kepler's laws

In 1601 Tycho died, and Kepler set to work with his observations to find an explanation of the planets' motions. After some years he concluded that all could be explained if the Sun were at the centre and the planets moved round it in orbits which were not circular but elliptical; in other words, if the distances of planets from the Sun varied from point to point in their orbits. Thus Kepler not only verified the heliocentric theory but also destroyed the Aristotelian notion that there was something of cosmic significance in the circle. He drew up three fundamental laws of planetary motion:

1. Planets move around the Sun in ellipses, with the Sun at one focus of the ellipse.
2. The line (radius vector) joining a planet to the Sun sweeps out equal areas of space in equal times. Thus a planet moves fastest when nearest the Sun and slowest when farthest away.
3. The square of the time taken to complete an orbit depends on the cube of the planet's distance from the Sun. Since the orbital periods of the planets were known, Kepler at once established the ratio of the distances of the planets from the Sun. Taking the Earth-Sun distance as one *astronomical unit*, a planet orbiting in 8 years would lie 4 astronomical units from the Sun ($8^2 = 64$, $64 = 4^3$).

The telescope and Galileo

The earliest use of the astronomical telescope was in 1609. Although there is some dispute as to who originally turned a telescope towards the heavens, there is no doubt that Galileo Galilei, an Italian, was the first man to do so systematically, and to appreciate the significance of what he saw.

While he was Professor of Mathematics at Padua, he

heard of the invention of the telescope by the Dutch spectacle-maker Lippershey, and proceeded to make one for himself. Although a tiny instrument, Galileo's telescope was powerful enough to show mountains and craters on the Moon, spots on the Sun, and the moons of Jupiter revolving round the planet just like a miniature solar system. He showed that the Milky Way consisted of myriads of stars and made many other previously inconceivable discoveries. From his observations he became convinced of the Copernican view of the universe and he was tried by the Inquisition for his 'heretical' ideas, and made to publicly denounce this hypothesis. He lived under 'house arrest' the rest of his life.

Newton
In 1642 Galileo died. In the same year was born the great British scientist Isaac Newton. He designed a completely new type of telescope – the *reflector* – which used a mirror instead of a lens to collect light. Newton's first reflector had a mirror 2·5 cm in diameter; the largest present-day reflector has a mirror of 6 metres.

Kepler had shown *how* the planets move – Newton showed *why* they move as they do. The story is told, and it seems to be true, that Newton once saw an apple fall from a tree and began to wonder why it fell. His chain of reasoning led him to propose the idea of universal gravitation. He showed that

The old Royal Observatory, Greenwich

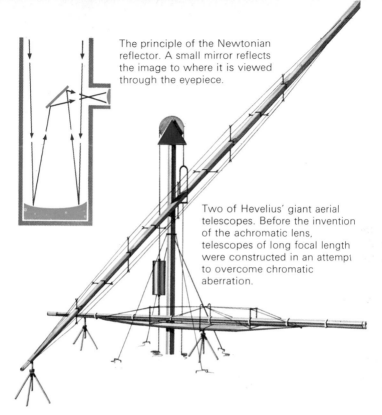

The principle of the Newtonian reflector. A small mirror reflects the image to where it is viewed through the eyepiece.

Two of Hevelius' giant aerial telescopes. Before the invention of the achromatic lens, telescopes of long focal length were constructed in an attempt to overcome chromatic aberration.

every object attracted every other object with a force depending on the masses of the objects and the inverse square of their distances apart: if the distance were doubled, the force was quartered. This force, too small to be noticed between everyday objects, is readily apparent with astronomical bodies. Gravitation keeps the fast-moving planets chained to their orbits instead of flying away from the Sun.

Progress speeds up

After Newton, developments were rapid, and it is only possible to list a few highlights of the subsequent 250 years. The year 1671 saw the foundation of the Paris Observatory, and 1675 the setting up of the Royal Observatory at Greenwich, through which it was decided that the zero line for

The Veil nebula

A typical spiral galaxy

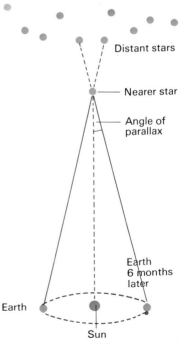

Distant stars

Nearer star

Angle of parallax

Earth 6 months later

Earth

Sun

longitude measurements on the Earth should run. Time was measured by checking the instants at which certain stars passed due south, or *transited*, and Greenwich became the centre of the world's time-keeping system.

Sir William Herschel, one of the greatest observational astronomers, discovered the planet Uranus beyond Saturn in 1781, and with his self-made telescopes, the best of their time, made observations of the distribution of stars in our stellar system, discovering the motion of the Sun in space in the process.

In 1838 Bessel measured the distance to the star 61 Cygni by the method of parallax – the first stellar distance ever measured. (Half the small angular difference in the star's position at six-monthly intervals is its *parallax*. Knowing this, trigonometry gives the star's distance.)

The planet Neptune, even more remote than Uranus, was discovered in 1846 after its prediction by Adams and Leverrier from mathematical calculations.

Huggins in 1863 identified elements in some stars by means of the spectrometer (page 28).

The famous 2·5 metre reflector was set up on Mount Wilson, California, in 1917 and brought many new discoveries. It was joined in 1948 by the 5·1 metre telescope on Palomar Mountain.

The last planet known in the solar system, Pluto, was found in 1930, pursuing a lonely path in the depths of space.

Since 1945, many new tools have been placed in the hands of astronomers: radio telescopes, for example. The present-day astronomer makes use of a host of techniques inconceivable twenty years ago. We now know that our Sun is a star amongst a hundred thousand million (10^{11}) in a galaxy which also contains clouds of gas and dust. Astronomers can trace the life-cycles of stars and galaxies, and speculate on the history and origin of the universe. We have come a long way from the primitive concepts of the Chinese and Egyptians.

The method of parallax. As the Earth moves round its orbit, a nearer star will appear to shift against the background of more distant stars. As the radius of the Earth's orbit is known, trigonometry will give the star's distance.

ASTRONOMICAL INSTRUMENTS

The equipment available to the present-day astronomer is varied. Even so, most astronomical research centres on the optical telescope, either the refractor or the reflector, used in conjunction with photography.

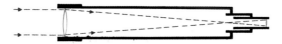

A simple Keplerian refracting telescope

In the simple lens light of different colours is refracted by different amounts, producing a fuzzy image with coloured fringes. This problem can be partially overcome by using a compound lens (*bottom*) made of two different kinds of glass with differing optical properties so as to cancel out the false colour effect.

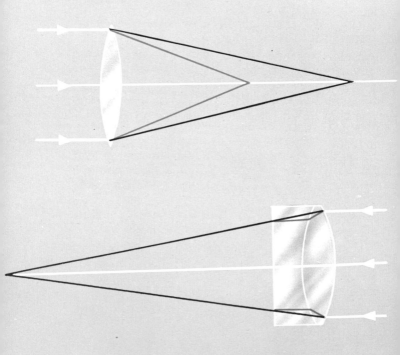

The refractor

As we saw, the refracting telescope was the form invented by Lippershey and used by Galileo and the early optical astronomers. Basically this telescope consists of two lenses, the convex *object glass* and the *eyepiece*. A convex lens will focus an image of a distant object on to a screen held at a certain distance from the lens. This distance is known as the focal length, and the ratio of this length to the diameter of the lens is the focal ratio; for example, a lens 5 cm across with a focal length of 61 cm will have a focal ratio of twelve, written as f/12. In a refractor the object glass is of large diameter and long focal length (with large 'f' ratio) and is used to gather a large quantity of light from the object in view and concentrate this light into an image. This image is then viewed by the eyepiece. In the case of the Galilean telescope the eyepiece lens is concave and produces an erect image. In the Keplerian telescope the eyepiece lens is convex and produces an inverted image. The Keplerian is the more satisfactory instrument. As we shall see (page 146), it is easy to make a refractor like this yourself in an evening.

Unfortunately the early refractors suffered from one serious disadvantage – their images were rather blurred, with coloured fringes around them. The reason for this is that, just as a prism splits light into its component colours, so a simple lens does not focus all colours to the same point. This phenomenon, known as *chromatic aberration*, can be partially overcome by making the object glass of very long focal length indeed. This led to the construction of the huge *aerial telescopes* such as that of Hevelius which was no less than 45 metres long. These instruments were extremely unwieldy and very hard to keep in focus. Seeing no solution to this problem, Newton was prompted to design the reflecting telescope.

However, in 1729 an English amateur, Chester More Hall, found that by assembling compound lenses, with two or more components made of different types of glass cemented together, he could more or less cancel out the false colour effects. This type of lens is the *achromatic lens*. The largest achromatic lens in existence is the 1·1 metre lens of the Yerkes refractor in the USA.

The reflector

Although the idea behind the reflecting telescope had been suggested some years earlier by the Scottish mathematician Gregory, the first working reflector was that presented by Newton in 1672. The principle of the reflector is quite simple: light is reflected from a concave mirror towards its focal point, where the image produced is viewed – in this case also by an eyepiece. The reflector has several advantages. For example, it does not suffer at all from the false-colour troubles of the refractor, and it is much easier to make and mount a mirror than a lens. Since light does not have to travel through the mirror, the quality of the glass is not so critical. Also, the mirror can be supported from behind.

The one complication with a reflector is that if the eye-piece is simply placed at the focus of the mirror, then anyone who tries to look through the lens will block out the light from whatever he is trying to observe. Newton's solution to this was to place a small flat mirror tilted at 45 degrees just before the focus, so that the light was reflected to the side of the tube where the image could then be observed without any difficulty. This type of reflector is known as the *Newtonian* and is still very much in use today, particularly amongst amateur observers.

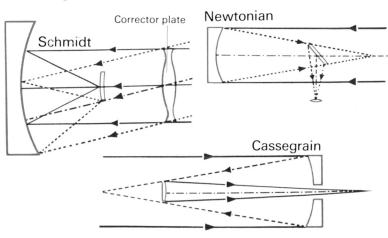

Herschel's famous reflecting telescope with 12 metre tube

Another common reflector is the *Cassegrain*. Here light from the main mirror is reflected back down the tube and through a hole in the main mirror to the eyepiece. With the Cassegrain reflector a long focal length can be achieved with a short tube, and the observer looks in the direction of the object he is observing (as with the refractor) instead of looking across the telescope tube as he does with the Newtonian.

The circle of sky seen in the eyepiece is known as the field of view, and with most telescopes this is very small. Since astronomers wish to photograph large regions of the sky at one time, other types of telescope have been evolved, the best known of these being the *Schmidt* reflector. Besides having a large, short-focus mirror, the Schmidt employs a glass correcting plate to remove various distortions in the large field of view.

Telescope mountings

The function of a telescope mounting is twofold. Firstly, of course, the mounting must hold the telescope rock-steady in all conditions so that the highly magnified image can be viewed or photographed. Secondly, there must be some simple means of moving the telescope so that it can follow the star in view in its apparent motion across the sky due to the Earth's rotation.

The simplest mounting is the *altazimuth*. The telescope is mounted on two axes, one of which is vertical. Rotation of the telescope about this axis will sweep parallel to the horizon. The other axis is horizontal enabling the telescope to be raised and lowered in altitude. Unfortunately, everywhere on Earth except at the poles the stars seem to move in paths which are not parallel to the horizon, and so it is rather difficult with the altazimuth mount to follow these motions. But if we move the vertical axis until it is parallel to the axis of the Earth, and fix it at that angle, then rotation of the telescope about this axis – the polar axis – alone will enable the observer to follow a star. This type of mounting is known as the *equatorial mount* and is used by all large telescopes. Usually some sort of motor is coupled to the mounting to drive the telescope round the polar axis at the exact speed of the Earth's rotation, so that the apparent motion of the stars is cancelled out.

Great telescopes

The largest refractors in use are the Yerkes 1·1 metre and the Lick Observatory 0·9 metre in the USA. The Meudon Observatory, near Paris, boasts a 0·8 metre refractor. Indeed, a 1·2 metre object glass was cast in France; but it was far from perfect and was never put into use.

Most of the largest reflectors are concentrated in the USA, the best known being the 5·1 metre reflector on Palomar Mountain in California, the 2·5 metre on nearby Mount Wilson, and the recent 3 metre at the Lick Observatory. The USSR has a 2·6 metre telescope in the Crimea and a new 6 metre telescope. Britain's largest telescope, the 2·5 metre Isaac Newton Telescope, is in operation on the island of La Palma, in the Canaries.

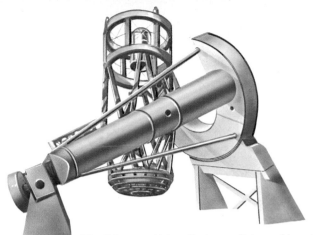

The 5·1 metre Hale reflector on Palomar Mountain

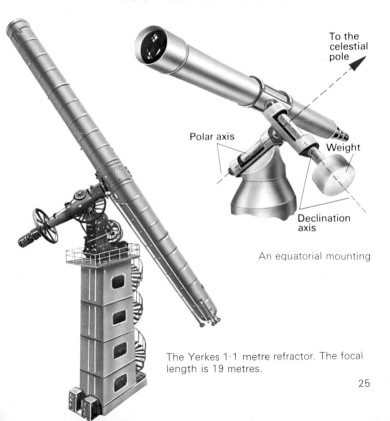

To the celestial pole

Polar axis

Weight

Declination axis

An equatorial mounting

The Yerkes 1·1 metre refractor. The focal length is 19 metres.

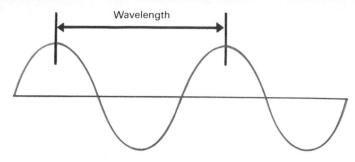
Wavelength

What is light?

The only way in which the optical astronomer can find out anything about the stars is by studying the light he receives from them. Light is a form of what is called electromagnetic radiation and can be regarded as a wave motion, rather like a water wave, with crests and troughs. The distance from crest to crest of a light wave is known as the wavelength, and visible light comes in a range of different wavelengths corresponding to different colours. In order of decreasing wavelength we have red, orange, yellow, green, blue, indigo, violet, corresponding to wavelengths from 7,500 Ångström units for red to 3,900 Å for violet. The Ångström is one hundred-millionth (10^{-8}) of a centimetre.

Longer wavelength light than red is known as infra-red; light of shorter wavelength than violet is ultra-violet. Neither is visible to the human eye.

If light enters a piece of glass at an angle, it is bent or 'refracted' out of its original direction. The shorter the wavelength, the more the light is refracted. Thus blue light, for example, is bent much more than red. White light, which we receive from the Sun, is a mixture of all the colours and so, if we pass white light through a glass prism, the various

The dark lines (Fraunhofer lines) crossing the spectrum indicate the wavelengths of light absorbed by cooler gases. Each element has a characteristic pattern or spectrogram — the spectrogram of a star or the Sun is a combination of the patterns of the elements present.

colours will be refracted by different amounts and the emerging light will be spread out into a band of colours like a rainbow, ranging from red to violet. This band of colours is known as a spectrum.

In 1814 the German physicist Fraunhofer found that the spectrum of the Sun was crossed by many dark lines, but was unable to account for them. Gustaff Kirchhoff showed that these dark lines were due to the absorbing effect of various elements in the Sun. All matter in the universe is made up of a combination of ninety-two basic materials or elements, and each element when heated enough emits light at certain characteristic wavelengths only. If these elements are in gaseous form, they can absorb light coming from behind at these same wavelengths. Thus, with the Sun, white light coming from the very hot lower regions, when passing through the cooler outer regions of its atmosphere, loses some of its light by absorption at certain wavelengths corresponding to the elements present.

It is thus possible to deduce what materials are present in the Sun by measuring the wavelengths of the dark lines, and comparing these with the spectra of elements in the laboratory.

A prism breaks up white light into its component colours.

The spectroscope

The instrument which astronomers use to analyze the light from the Sun and stars is known as the spectroscope. In its simplest form the instrument consists of a slit through which light from the star enters; the collimator, a lens which focuses the light into a parallel beam; a prism which splits up the star's light; and a small telescope which forms an image of the spectrum so obtained. More often the telescope focuses the spectrum on to a photographic plate, and the instrument is then more properly called a *spectrograph*. The photograph so obtained is a *spectrogram*. With stellar spectra, very long time exposures with the camera are needed because the light is so feeble. The mounting and driving mechanism of the telescope on which the spectroscope is mounted must be absolutely accurate; otherwise a blurred and useless spectrogram will result.

Sometimes the prism is replaced by a *diffraction grating*, which is simply a plate ruled with thousands of closely spaced lines. The diffraction grating performs the same function as the prism.

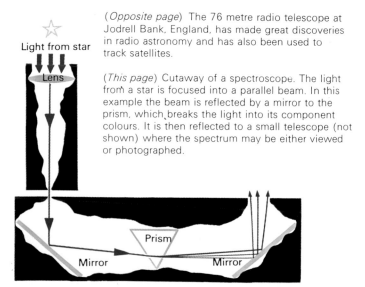

(*Opposite page*) The 76 metre radio telescope at Jodrell Bank, England, has made great discoveries in radio astronomy and has also been used to track satellites.

(*This page*) Cutaway of a spectroscope. The light from a star is focused into a parallel beam. In this example the beam is reflected by a mirror to the prism, which breaks the light into its component colours. It is then reflected to a small telescope (not shown) where the spectrum may be either viewed or photographed.

Light from star

Lens

Prism

Mirror

Mirror

Radio astronomy

Karl Jansky, an American physicist doing research into the problems of radio communications, discovered as long ago as 1931 the existence of radio waves coming from the direction of the centre of our galaxy, but it is only in the last twenty years that the science of radio astronomy has emerged as being of the utmost importance. Radio waves, like light, are simply a form of electromagnetic radiation. The only thing that distinguishes them from visible light is their wavelength, for radio waves range from about a centimetre upwards, being at least ten thousand times longer than light waves.

This radiation from space is very feeble indeed and enormous telescopes are required to gather enough radiation to give a detectable signal. One of the most famous radio telescopes in the world is of course the 76 metre steerable 'dish-type' instrument at Jodrell Bank in England. This type of radio telescope is quite obviously analogous to the reflecting optical telescope, focusing radio waves on to an aerial suspended in front of the dish.

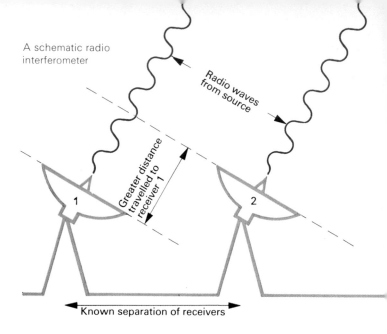

A schematic radio interferometer

Radio waves from source

Greater distance travelled to receiver 1

1

2

Known separation of receivers

However, other types of radio telescope bear no such obvious resemblance, and are used for specialized purposes. A word should be said about the *radio interferometer*. When light or radio waves meet, they are said to *interfere*. If the crests of two waves meet, the effect is to produce a more intense wave and the two waves are said to interfere constructively; should the crest of one coincide with the trough of the other, then the two waves cancel out and the waves are said to interfere destructively. When two beams of light are mixed together, we get patterns of constructive and destructive interference. Now, if we have two radio aerials set a considerable distance apart, then the distance that a radio beam has to travel from the source to the first aerial will not be quite the same as to the second aerial. Thus when one aerial is receiving, say, a crest, the other may not, and if we combine the signals from both aerials, we shall get formation of interference patterns. By manipulation of the two aerials and the resulting signals we can determine very accurately the position in the sky of a radio source.

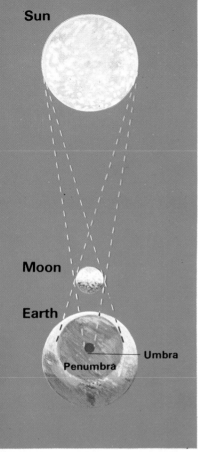

Sun

Moon

Earth

Umbra

Penumbra

When the Moon passes in front of the Sun, it blocks out the Sun's disc from view, causing an eclipse of the Sun. Likewise, if the Moon passes through the Earth's shadow, the Sun's light is cut off, and an eclipse of the Moon occurs. The Moon does not completely vanish during a lunar eclipse because the Earth's atmosphere acts just like a lens and refracts the Sun's light, so that some still reaches the Moon.

Why does an eclipse of the Sun and of the Moon not occur every month at New Moon and Full Moon? The reason is that the plane of the Moon's path is not quite the same as the plane in which the Earth moves: consequently, at New Moon, the Moon is more likely to pass above or below the Sun than actually to pass in front of it. Likewise, the Moon is more likely to pass near the shadow of the Earth than go through its shadow.

When the Sun is completely obscured by the

(*Top*) When a solar eclipse occurs, a total eclipse is seen from a small area, the umbra, and a partial eclipse from a larger area, the penumbra. (*Below*) Total eclipse of the Sun

tion), and again when they are on the opposite side of the Sun (*superior conjunction*). The outer planets (known as the *superior planets*) can only come into superior conjunction, for they do not pass between the Earth and the Sun. When an outer planet is opposite the Sun in the sky, that is, when Sun, Earth and planet are in line with the Earth in the middle, the planet is said to be in *opposition*. It is then at its nearest and appears brightest. When an inner planet is at its farthest point in the sky from the Sun, it is said to be at *greatest elongation*.

The Moon travels round the Earth at an average distance of 386,000 km in a period of about 27 days. During this time it goes through a complete cycle of phases, as shown in the illustration below. The Moon, like the planets, has no light of its own and shines only by reflecting sunlight. When the Moon is in line with the Sun it cannot be seen. As it moves away from the direction of the Sun, we can see more and more of the illuminated side until it is opposite the Sun in the sky at Full Moon and the whole orb is visible. The Moon then begins to approach the position of the Sun once again and shrinks back through a thin crescent to New Moon.

(*Inset*) The phases of the Moon are shown together with the corresponding positions of the Moon in its orbit around the Earth.

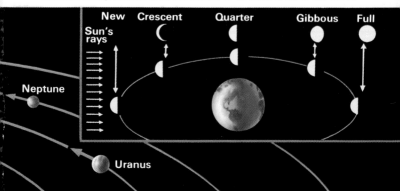

THE SOLAR SYSTEM

The Sun carries with it in its journey through space not only the nine planets, but a host of other minor bodies, such as comets, meteors, and the minor planets or asteroids. The planets are, in order from the Sun: Mercury, Venus, Earth, Mars, Jupiter, Saturn, Uranus, Neptune and Pluto. Although the planet Pluto is, on average, a hundred times farther away from the Sun than Mercury, it is not always the outermost planet: the eccentricity of its orbit is so great that it actually comes closer in than Neptune for about a sixth of its 248-year orbital period. Between the orbits of Mars and Jupiter lie the thousands of minor planets whose origin is rather a mystery.

Comets and showers of meteors tend to move in highly eccentric ellipses – in fact some comets move in orbits of such a kind that they may never return again from the depths of space.

One or two technical terms concerned with planetary orbits are useful to know. A planet is said to be in *conjunction* when it is in line with the Sun as seen from the Earth. The inner two planets, Mercury and Venus, come into conjunction at two points in their paths – once when they pass between the Sun and the Earth (known as *inferior conjunc-*

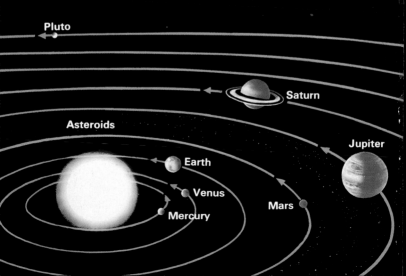

Other techniques

Our atmosphere will let only radiation of certain wavelengths pass through: in fact, only visible light and some radio waves. This is extremely fortunate from our point of view, as many of the other radiations coming from space are very harmful indeed to human beings. However, from the astronomer's point of view it is rather a nuisance, as he would like to study these other radiations to see what he could find out from them. Furthermore, our atmosphere does not even let visible light through all that well. Apart from the obvious problem of clouds, the atmosphere is dusty and turbulent, and these factors combine to make observing very difficult from ground level.

One way to reduce this difficulty is to build observatories as high up as possible. This cuts down some of the atmospheric effects. But the only real solution is to get clear of the atmosphere completely by means of rockets, satellites and space probes. An observatory on the Moon, which has no atmosphere, would enable astronomers to make observations unimpeded by dust and clouds. This should be possible in the not-too-distant future.

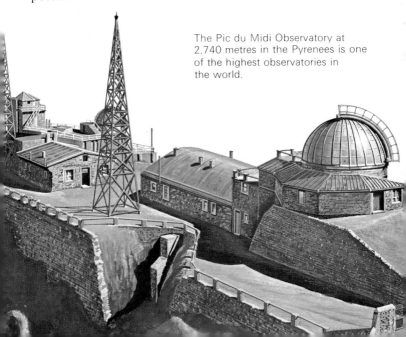

The Pic du Midi Observatory at 2,740 metres in the Pyrenees is one of the highest observatories in the world.

Transit of Venus

Moon, we have a *total eclipse* of the Sun. Total eclipses are fairly rare and are of great interest to astronomers, so that expeditions are sent to the places from where they can be best observed whenever they occur. As the distance from the Earth to the Moon varies slightly, the Moon sometimes appears smaller than the Sun, and an *annular eclipse* can occur, when the bright rim of the Sun is still visible.

It is more common for the Moon to cover only part of the Sun, when a *partial eclipse* is said to occur. Partial and total eclipses of the Moon also occur.

Mercury and Venus can also pass in front of the Sun, such events being known as *transits*. These occurrences are very rare indeed, and used to be of great importance since, if two observers a long way apart observe the exact instant at which the planet just appears on the face of the Sun, they will not get precisely the same result on account of parallax (see page 18). Thus the planet's distance can be found. Kepler's Laws will then give the distances of all the other planets. Nowadays it is possible to measure the distance of Venus by radar.

The origin of the solar system

A problem that has concerned astronomers for generations is the question of how the planets and other bodies of the solar system were formed. An early scientific theory was the *nebular hypothesis* of the French mathematician Laplace (1749–1827). Laplace supposed that the Sun was formed from a gas cloud which, as it contracted and grew hotter, began to rotate at high speed. The Sun began to bulge at its equator until finally it threw off a ring of material. This process was repeated several times and it was supposed that these rings of matter would condense into planets. However, it has been shown that such rings – even if they could be formed – would not condense into planets.

Sir James Jeans, in the early part of the twentieth century, proposed that early in the Sun's history another star passed close by, and the gravitational interaction between the Sun and the star drew a cigar-shaped filament of matter from the Sun. As the other star receded, it set this matter in motion around the Sun, the matter ultimately condensing into planets. This theory also superficially accounted for the differing sizes of the planets, since those nearer the middle of the 'cigar' would presumably be larger. However, once again the theory was not mathematically sound and was eventually abandoned.

Since 1950 there has been something of a return to the

According to Sir James Jeans' theory of the origin of the solar system, a passing star drew out from the Sun a filament of material which condensed to form the planets.

Second star

Material from Sun

Sun

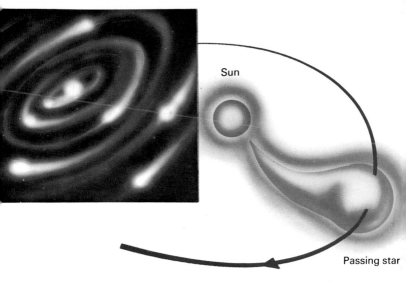

Sun

Passing star

(*Left*) The nebular hypothesis of Laplace suggested that the Sun threw off rings of material which condensed to form planets. (*Right*) M. M. Woolfson has suggested that the Sun attracted material from a passing star. This material went into motion around the Sun and condensed to form the planets.

nebular hypothesis, and several theories have been proposed, based on a similar general theme. It is supposed that the Sun was surrounded by a cloud of gas and dust. It is thought that this material would settle into a disc and that irregularities would cause accretion to commence; that is, lumps of material would begin to form. Large lumps would tend to attract other pieces of matter, and it is thought that in this way the planets might have condensed. In the last few years, Professor M. M. Woolfson of York University has proposed an interesting variation of the hypothesis of Jeans. Woolfson suggests that, instead of material being drawn from the Sun, it was in fact drawn from the passing star. This would overcome many of the objections to Jeans' original hypothesis. However, at the present time there is no really satisfactory solution to the problem of the origin of the solar system.

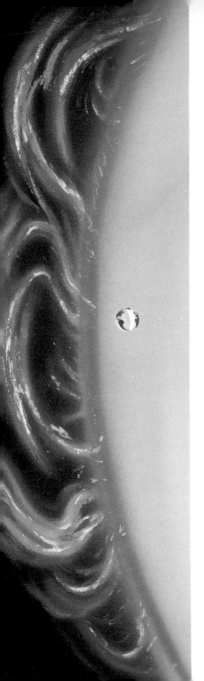

THE SUN

The Sun is the nearest star to the Earth, and a very ordinary star at that; although to us it is of vital importance, it is only one of thousands of millions of such bodies within our galaxy. Living on our very solid Earth, it is hard to realize that we are hurtling round the Sun at something like seventy thousand miles an hour! It is equally hard to appreciate under what precarious circumstances we exist. Were the total amount of solar radiation to change by only a small fraction, then human life as we know it would be impossible – we would either freeze or fry. In fact, it is thought by many people that the ice ages in the past were due to very slight changes in the Sun's output of energy. Fortunately for us the Sun is a very stable star and not one of the pulsating or exploding stars that we shall meet later (page 84).

The Sun's diameter of about 1,390,000 km is enormous by comparison with the Earth's 12,700 km. The

The Earth in comparison with the Sun

surface temperature of the Sun is about 6,000 Kelvin (K) (the Kelvin, or Absolute, temperature scale has its zero at $-273°$ Celcius), while near the centre the temperature is thought to be about 20,000,000 K.

Although the volume of the Sun is about 1,300,000 times that of the Earth, its mass is only about 330,000 times greater, and so we can see that the Sun has a much lower density than the Earth. The average density of the Sun is about 1·4 times that of water, whereas the Earth's density is some 5·5 times that of water. The Sun is a completely gaseous body; near the centre the gas is very highly compressed and extremely dense, while the outer regions are tenuous and of very low density.

The gaseous Sun does not rotate like a solid body: different regions rotate at different velocities. The material near the Sun's equator revolves in about 25 days, whereas at 40° north (or south) the period is $27\frac{1}{2}$ days, and at the poles some 34 days. This differential rotation of the Sun is easily seen in the motions of dark spots – sunspots – on the surface.

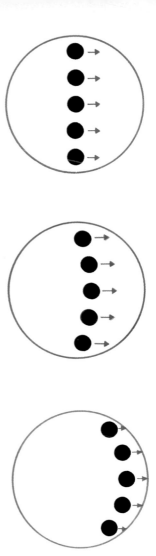

Observation of sunspots shows that the equatorial regions of the Sun revolve faster than the polar regions.

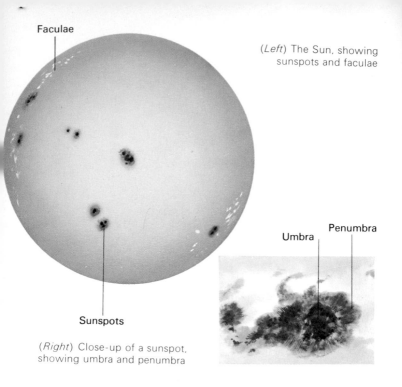

Faculae

(*Left*) The Sun, showing sunspots and faculae

Umbra Penumbra

Sunspots

(*Right*) Close-up of a sunspot, showing umbra and penumbra

The surface of the Sun which we see is known as the *photosphere* ('sphere of light') and it is in this region that the sunspots appear. These spots were first seen telescopically by Galileo, but there is good reason to believe that the early Chinese knew of their existence, as occasionally some are large enough to be visible to the naked eye when the Sun is dimmed enough – for example, by haze near the horizon. With a small telescope, such sunspots are readily visible (*but* see page 148).

The structure of a sunspot is often quite complex. All sunspots have a dark central region, the *umbra*, surrounded by a lighter outer area, the *penumbra*. The spots seem to be saucer-shaped. They are, in fact, regions which are cooler than the rest of the Sun's surface; the gases in these regions are at temperatures around 4,500 K. Thus sunspots only appear dark by contrast with the photosphere.

Strong magnetic fields are associated with the sunspots, but the reason for their existence is not fully understood. The number of spots varies periodically over what is known as the *sunspot cycle*. At the sunspot minimum, only 50 sunspot groups a year will be seen, but at maximum there may be 500 or more. This cycle extends over a period of about 11 years, and there may be longer-term variations as well. Large spots can, however, appear at any stage of the cycle. The colossal group which appeared in 1947 covered an area of 5,000 million square km.

Often associated with the sunspots are bright patches known as *faculae*, which seem to lie at a higher level than the spots. Very often, faculae will be seen just before a spot breaks out, and they will generally persist for a time after the spot itself has vanished.

The most dramatic of the solar features are the *prominences*, enormous bright patches which appear to shoot up like flames from the surface of the Sun into its rarefied outer atmosphere (the corona). Sometimes attaining lengths of hundreds of thousands of kilometres, these prominences are due as much to gases condensing out of the corona as to material shot up from the surface.

Solar prominences appear to shoot out of the Sun like flames.

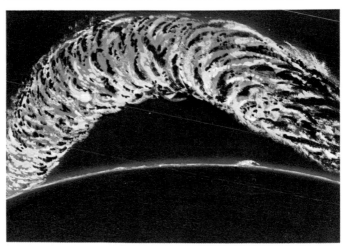

A solar telescope

Apart from daily counts of sunspot numbers and general features, fairly sophisticated equipment is needed for the study of the Sun, and many special solar observatories are in existence. Most solar observatories such as the Mount Wilson in the USA and Arcetri in Italy have rather odd-looking tower telescopes which are used to produce stable images of the Sun. As solar astronomers must, of course, work during the day, they have to contend with heat and hot air currents, and to help overcome this problem the Sun's image is reflected down through the tower to an underground room, which is kept at a constant temperature. Here the image is projected on to a screen, or analyzed by means of spectroscopes and similar equipment, such as the *spectroheliograph*. This is an extremely useful piece of equipment which enables the observer to view the Sun in light of one particular wavelength. Thus he can choose, for example, the wavelength of the Fraunhofer lines corresponding to

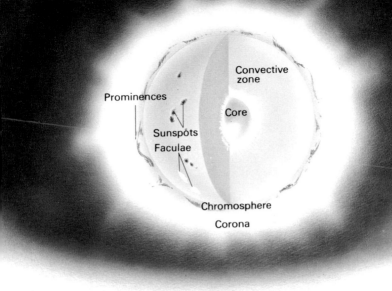

Prominences

Convective zone

Core

Sunspots

Faculae

Chromosphere

Corona

A cutaway section of the Sun

the element calcium or hydrogen, and study the surface distribution of this material. Also of value is the *Lyot monochromatic filter*, which provides a more convenient method of studying the Sun at particular wavelengths and showing the prominences. The inner corona may be studied without waiting for a total eclipse by means of the *coronagraph*.

The Sun has quite a complex structure. As it is a ball of gas, there is difficulty in defining what the 'surface' is, or deciding where it really is. We can divide up the Sun into several layers. In the centre is the *core* where all the tremendous energy is generated, and outside this is a broad zone which merely serves to transport heat to the surface in the form of radiation. The surface which we see with the naked eye is the photosphere, but beyond this is a further layer, the *chromosphere*, only about 4,830 km thick, which is only visible at a total eclipse (when it appears to be pink in colour), except by means of the spectroheliograph. The spectral lines from this zone are bright emission lines, as distinct from the Fraunhofer absorption lines, but are normally too faint to be seen in the solar spectrum. Just as the Sun goes into eclipse, these lines flash briefly into sight,

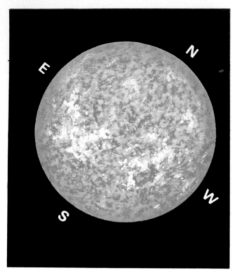

A spectroheliogram. In the spectroheliograph the image of the Sun passes through a prism and a slit whose position can be adjusted to pass light of any required colour. By moving the image of the Sun across the slit in synchronization with moving a photographic plate across the slit, a photograph of the Sun is obtained in light of the desired wavelength.

giving rise to the 'flash spectrum' which tells us what materials are present. The spectroheliograph enables us to study the fascinating structure of this area.

Beyond the chromosphere lies the corona, which again is normally invisible except at a total eclipse, when it surrounds the dark disc of the Moon with a pearly halo which varies in size and structure with the eleven-year sunspot cycle. The gas in this region is very rarefied indeed – which is as well for us, as the atoms are hurtling round at speeds corresponding to temperatures of about a million degrees, and if the whole Sun were radiating at this temperature, then it would be very hot here on Earth.

We know that the Sun is a ball of gas; but of which gases is it composed? Analysis of the solar spectrum shows us that the principal one is hydrogen, which is relatively rare in our own atmosphere, but is thought to compose about 70 per cent or 80 per cent of the mass of the Sun. Another major component is helium, which is the next lightest gas (hydrogen being the lightest of all the elements). It is rather interesting to learn that helium was first discovered, not on Earth, but in the solar spectrum. In 1869, the English astronomer Lockyer found a Fraunhofer line that he could not identify as being due to any known element, and he supposed – correctly – that this line was due to some unknown element which he called helium after the Greek word for 'Sun'. It was not until a quarter of a century later that helium was found on Earth. Many other materials have since been identified in the solar spectrum, the most common, apart from hydrogen and helium, being carbon, nitrogen, oxygen, and the vaporized metals magnesium and iron. At least 70 of the 92 natural elements are now known to exist in the Sun, and we presume that the others do too.

Where does the Sun get its vast resources of energy? After all, it has been shining for thousands of millions of years, and should continue to do so for thousands of millions more. Were it simply burning as coal burns, then it would have gone out long before now. When atoms of one element are joined together to form another element, in the process known as *fusion*, tremendous amounts of energy are released, and this is what goes on in the centre of the Sun. Basically what happens is that atoms of hydrogen (which is in plentiful supply) are fused together at temperatures in excess of 14,000,000°C to form helium, with the release of energy. The hydrogen bomb uses just this principle, so you can regard the Sun as a controlled hydrogen bomb if you wish.

The actual process is thought to occur in three stages as follows. Firstly, the nuclei of two hydrogen atoms are joined together to form a deuteron, or 'heavy hydrogen', nucleus, with the release of a positron, a particle like the electron but with positive charge. The deuteron then combines with another hydrogen nucleus (or proton, in other words) to form an isotope of helium with atomic weight 3, together with the release of energy in the form of short-wavelength gamma radiation. Finally two helium 3 isotopes combine to form one ordinary helium (atomic weight 4) nucleus with the release of two protons which are then free to take part in the process again. This process of synthesis of helium in the Sun's core is known as the proton-proton reaction. Many stars are thought to obtain energy in this way.

The Sun emits radiation at a whole range of wavelengths besides the visible range of light. In fact, it is known that the Sun emits all kinds of radiation, from the very short wavelength 'hard' X-rays (which would be deadly to human life if they were to penetrate the atmosphere) to radio waves, some of which can be studied from the surface of the Earth. The X-rays can be studied by means of rockets. The total amount of radiation reaching the Earth is termed the *solar constant* and is found to be 1390 joules/square metre/second – in other words, enough heats lands per minute on a square centimetre of the atmosphere to heat 1 gram of water by 1·95° Celcius (3·5° Fahrenheit), and this amount of radiation remains remarkably constant, making life here possible.

The Sun also sends out streams of various particles, such as atoms, the central parts of helium atoms (known as nuclei), and electrons and other electrically charged particles, particularly after the appearance of sunspots and other such storms on the surface of the Sun. Many of these particles become trapped by the magnetic field of the Earth.

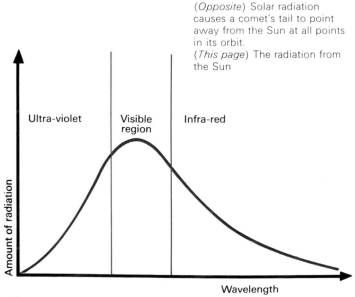

(*Opposite*) Solar radiation causes a comet's tail to point away from the Sun at all points in its orbit.
(*This page*) The radiation from the Sun

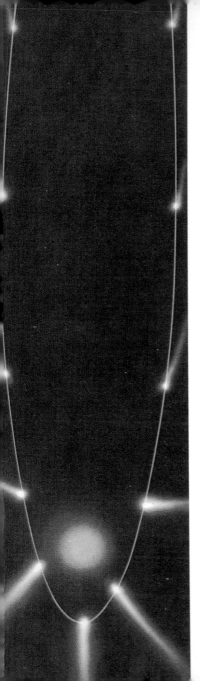

They give rise to radiation belts above the atmosphere and cause such phenomena as auroral displays and radio interference. The stream of particles blown out from the Sun is known, appropriately, as the *solar wind*. Thus some people have contended that the Earth moves in the very outer regions of the Sun's atmosphere or corona, and so we see again that the boundary of the Sun is very hard to define.

The birth and death of the Sun

The Sun is thought to have formed from a huge cloud of gas and dust, of which there are many in our galaxy. Slowly the gas cloud contracted, and as it did so it began to get hotter and to spin. Eventually the pressure and temperature in the centre became sufficient for nuclear reactions to begin, and the Sun rapidly attained its present temperature and settled down to a stable state. Ultimately, though, the Sun will use up all its nuclear fuel, contract, and become a tiny white dwarf star – but this need not concern us much, as it will take place at the earliest in five thousand million years' time.

47

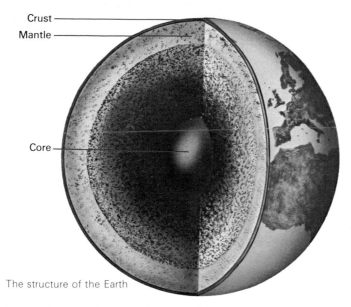

Crust
Mantle
Core

The structure of the Earth

THE PLANETS AND THEIR SATELLITES
Earth

The Earth is, as yet, the only planet of which we have direct experience, and although we know it to be an insignificant body in the universe, it nevertheless seems very solid and important to us. What then do we know about our planet?

The Earth revolves around the Sun at an average distance of about 149 million km in a period of 365¼ days. The orbit is slightly elliptical, the Earth being at its nearest to the Sun, and thus moving at its fastest, during the southern hemisphere summer, and farthest away during the northern hemisphere summer. This explains why summer in the southern hemisphere tends to be shorter and hotter than summer in the northern. The Earth is nearly spherical in shape, the diameter between poles being a few kilometres less than the equatorial diameter of 12,750 km. The Earth revolves on its axis once every 23 hours 56 minutes.

Largely by studying the shock waves caused by earthquakes, geophysicists have been able to deduce the

structure of the Earth's globe. In the centre lies the very dense *core* composed of highly compressed rocks and iron in a liquid state, while outside this and extending almost to the surface is the rocky *mantle*, the former being about 3,400 km in radius and the latter extending a further 2,900. On top of the mantle lies the *crust*, 50 to 65 km thick and composed of everyday rocks far less dense than those in the mantle. The crust is the material of which continents are composed. Finally, the few feet of soil and few miles of water on the surface support all life on Earth.

Surrounding the solid globe are the gases which make up our atmosphere, principally oxygen, nitrogen, argon, carbon dioxide, and water vapour, the most abundant being nitrogen. Oxygen is essential for human life, whereas plants absorb carbon dioxide and expel oxygen. The carbon dioxide also helps to trap the heat of the Sun and prevent it all escaping at night. The height of the atmosphere is hard to determine, but there are detect-

A section through the atmosphere

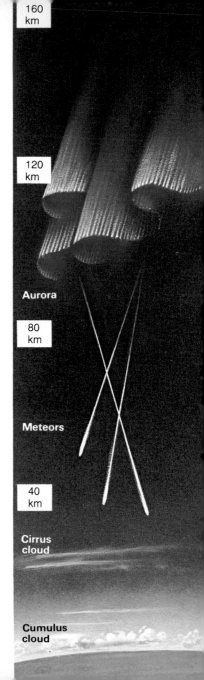

160 km

120 km

Aurora

80 km

Meteors

40 km

Cirrus cloud

Cumulus cloud

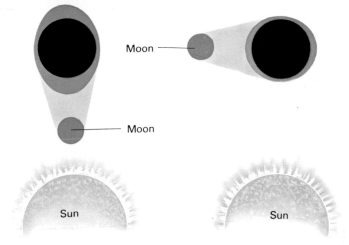

Spring tides occur (*left*) when the Sun and Moon are in line with the Earth and their attractive forces are added. Neap tides occur (*right*) when the Sun and Moon are at right angles and their pulls tend to cancel out.

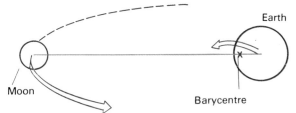

We speak of the Moon revolving around the Earth, but in fact both revolve around the barycentre (x) or centre of gravity of the Earth-Moon system.

The aurora

able traces at over 800 km, although half of the mass of the atmosphere lies within five km of sea level.

Various earthly phenomena have astronomical origins. The *tides*, for example, are due to the gravitational effects of the Sun and Moon. Basically, the water in the oceans is attracted by the Sun and Moon and tends to bulge up at the point where the Moon is overhead, while at the same time a lesser bulge is induced on the opposite side of the Earth, these bulges corresponding to high tide. As the Earth rotates once daily, everywhere has two high tides as the bulges pass by. Exceptionally high and low (Spring) tides occur when the Sun and Moon are in line and are pulling in the same direction, while the least extreme tides occur when Sun and Moon are separated by a right angle and their pulls tend to cancel out (Neap tides). Thus we can expect very high and low tides at New and Full Moon, and not nearly such high and low tides at first and last quarter.

The *aurorae* (northern or southern lights, depending on where you live) are due to the effects of charged particles from the Sun spiralling down the Earth's magnetic field. These very beautiful phenomena can take many forms, such as moving rays, hanging curtains, and bright arcs, and generally show a predominantly green hue with pinks and yellows on occasion. They seem to take place at heights from ninety to several hundred kilometres. Most displays can only be seen from polar regions, but the odd exceptional aurora can be seen even from the equator.

About 120 km up in the atmosphere comes the layer known as the *ionosphere*, a layer of charged particles which reflects many radio waves and makes long-distance radio communication possible by bouncing waves around the Earth. This layer is highly susceptible to disruption by charged particles from the Sun, especially after some solar outburst. This disruption gives rise to radio interference.

About this height, too, *meteors* begin to become visible. Often known as *shooting stars*, these are nothing more than tiny particles of matter (mostly about the size of sand grains) coming from space at high speed and burning up by friction in the atmosphere. Larger bodies, of different types, reach the Earth's surface and are known as *meteorites*.

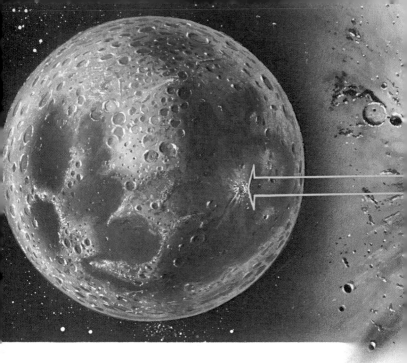

The Moon

Our Moon revolves around the Earth in a period of about 27 days; it also revolves once on its axis in the same time and so it always keeps the same face towards the Earth. This phenomenon, known as *captured rotation*, is due to the tremendous gravitational pull of the Earth on the Moon. As the Moon moves faster at one part of its orbit than another, we are able to see first a little way around one side and then a little way around the other, but nevertheless most of the far side is permanently hidden from the Earth. It was not until the Russians sent their spacecraft Lunik III round the far side in 1959 that we first received photographs of the hidden side. Since then much more has been learned about the far side, and the visible side too, particularly as a result of the American Lunar Orbiter satellites and of course the Apollo programme.

The Moon is a desolate, hostile world of barren rocks, towering mountains – some higher than Mount Everest –

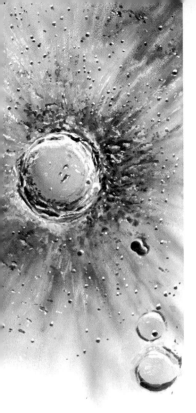

and huge plains. There is no air on the Moon, for its gravitational pull is too small to retain any gas. The Moon is 3,500 km in diameter. Its mass is only 1/81 that of the Earth, and the gravitational force at the surface is only 1/6 of the force at the surface of the Earth. Since it is this force which gives us our 'weight', an 80 kilogram (176 pound) astronaut will weigh 13 kilograms (29 pounds) on the Moon. The astronaut must be equipped for great extremes of temperature, as the temperature in the middle of the sunlit side is about 100°C, and on the dark side, −150°C (272° below freezing, Fahrenheit). The main surface features are the craters, which range from pits to huge depressions over 240 km across, the mountain ranges, some of which are 1,600 km long, and the relatively flat plains.

The Moon, pock-marked with craters and (enlarged) the crater Copernicus.

Mare Imbrium. Early astronomers thought such areas, which appeared flat, were seas. They are not in fact entirely flat but may have craters and mountains.

Lunar features – the crater Aristarchus
and (*inset*) the Straight Wall

Besides these major features there are many valleys,
gorges, ridges, faults, and chains of little craterlets. Systems
of bright rays spread out from the rims of some craters; these
these seem to be due to some sort of material somehow
expelled from these craters.

The origin of the craters has been a cause of scientific
dispute for a long time now, the two main theories that hold
sway being the *meteoritic* (impact) theory and the *volcanic*
(internal activity) theory. The former theory proposes that
the craters were formed by the explosive impact of meteorites hitting the Moon, mainly in its early history. There are,
indeed, meteorite craters on Earth (see page 74), but there
are volcanic ('vulcanistic') craters too. There is indeed some
evidence for lingering volcanic activity on the Moon at the
present time. The truth probably lies between the two
theories – there must be meteor craters and volcanic craters
on the Moon: the only problem is which was the major
crater-forming process. The US Apollo lunar missions went
some way towards answering these questions.

54

Mercury

Mercury is the smallest planet and the one nearest the Sun, dashing around in its orbit in a period of only 88 days. Radar measurements have shown that Mercury does not – as was previously thought – keep the same face permanently turned towards the Sun, but rotates on its axis in 59 days. The temperature in the middle of the sunny side reaches 400°C while on the dark side it drops below − 180°C.

The diameter of the planet is only 4,880 km and there seems to be a very thin atmosphere consisting of minute traces of helium, neon and argon. A more inhospitable place it is hard to imagine! Being an inner planet, Mercury shows phases, but it is unfortunately very hard to observe – because of its small size and proximity to the Sun. Only occasionally is it visible to the naked eye – just before sunrise or just after sunset when it is near greatest elongation.

The Mariner 10 spacecraft visited Mercury in 1974 and 1975. It sent back pictures showing a surface very much like that of the Moon.

Artist's impression of the surface of Mercury

Venus

Venus, named after the goddess of love, is a beautiful sight in the morning or evening sky, when she shines with a brilliant silvery hue. She is known by many as the *Morning Star* or *Evening Star*, as the case may be. At her brightest she is the brightest object in the sky besides the Sun and Moon.

Venus is the closest planet to the Earth in size, being 12,104 km in diameter, 652 km less than the Earth. Moving in an orbit 108 million km from the Sun, Venus can come as close to the Earth as 40 million km – much closer than any other planet. The mass of Venus is similar to the Earth's.

A 'day' on Venus is longer than the Venusian (or 'Cytherean') 'year'. This curious situation arises as Venus revolves faster around the Sun than it rotates on its axis, the former period being 224 days and the latter, 243 days. Thus Venus appears to rotate backwards with respect to the Sun, and so the Sun would appear from Venus to rise in the west and

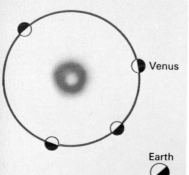

(*Above*) The phases of Venus
(*Left*) From this diagram it can be seen that Venus will be almost invisible when closest to Earth.

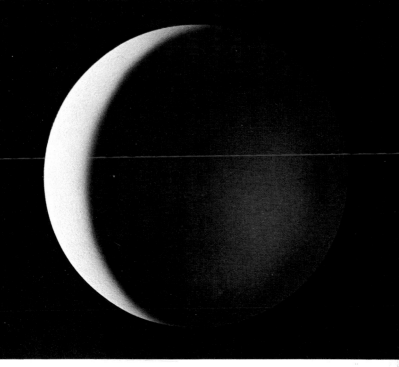

The 'ashen light' of Venus

set in the east. The entire planet is shrouded by thick carbon dioxide clouds which never allow astronomers to see the surface. But a number of unmanned spacecraft have visited Venus and have used highly accurate radar to 'see' through the clouds and 'map' the surface. Also, two Soviet probes have landed on the surface and sent back pictures of the hostile, rock strewn landscape.

The various probes have revealed a surface temperature of the order of 475°C and an atmospheric pressure about 90 times that of the Earth's. The atmosphere is highly corrosive, as droplets of sulphuric acid constantly rain down on the surface – it's no wonder that none of the landers has lasted more than a few hours.

Mars

The first, and probably best known, of the superior planets is Mars, named after the god of war because of its red colour. Mars revolves round the Sun at an average distance of 227,900,000 km in a period of 687 days. Although only 6,790 km in diameter, Mars is in many ways far more like Earth than Venus is. Mars is not covered by cloud and we can see clearly down to the surface, which has a predominantly reddish colour against which are displayed many more-or-less permanent dark patches of a blue-green tint, thought by early astronomers to be seas.

Another interesting feature of Mars is the presence of white *polar caps*. These appear to be rather thin but made up of water ice like the Earth's polar caps. In each hemisphere, as summer approaches, the polar cap shrinks, releasing a quantity of water vapour into the Martian atmosphere. The cap grows again with the onset of winter. Mars has always been considered the most likely abode of life apart from the Earth in the solar system, and this may prove to be the case. But human life could not survive there and it is doubtful whether even the lowest of earthly plants could either. For one thing, the atmosphere is very thin (equal to the Earth's at a height of 30 km) and is composed mainly of carbon dioxide with very little water vapour and practically no oxygen at all. Also, although the tempera-

The surface of Mars is now known to have craters. This picture is based on close-up photos by Mariner 4.

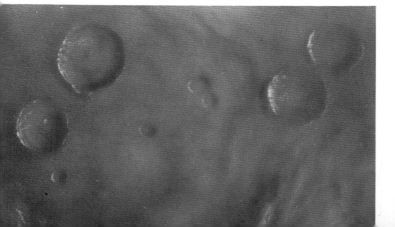

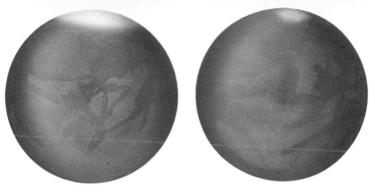

Mars, showing north polar cap in winter (*left*) and summer (*right*)

ture can reach −20°C (12°F) in some regions, the night temperature is very low, with a Martian day of 24 hours 37 minutes.

Photographs taken by the Mariner, Viking, Soviet Phobos 2 spacecraft show that Mars is covered with craters and valleys. A number of obviously volcanic features are also present. The surface consists of boulders and reddish dust, while the Martian sky appears pink due to the effect of suspended dust. Mars has two tiny moons, Phobos and Deimos, which are 22 and 12 km across respectively. No life form has been found.

Mars has two small moons, Phobos (to the left of the planet) and Deimos.

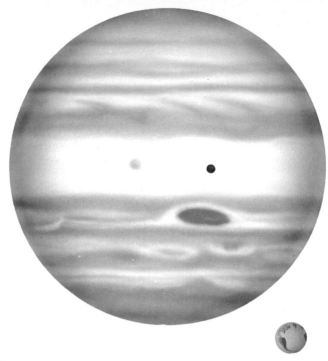

Jupiter, showing the Great Red Spot. The Earth is shown to the same scale for comparison of size.

Jupiter

Jupiter was the king of the gods and so his title befits the largest planet: Jupiter is no less than 142,800 km in diameter at the equator. However, as Jupiter rotates in the very short period of 9 hours 51 minutes, it bulges at the equator and is flattened at the poles, the polar diameter being 134,080 km. Jupiter moves slowly round its orbit, taking nearly twelve years to complete a circuit of the Sun at an average distance of 778 million km.

Jupiter is not a mainly solid body like the Earth but seems to be largely gaseous with the rather low density of only 1·3 times that of water. This gas seems to be principally helium, and associated gases such as methane and ammonia. Whether or not Jupiter has a solid core is a matter

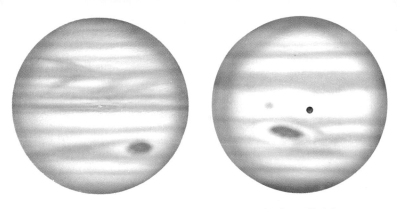

Observations taken at different times show that the Great Red Spot moves in the atmosphere of Jupiter.

of some debate, and it may be that the central parts are simply gas compressed to very high densities and behaving more like metal. The temperature is extremely low at the visible surface, about $-150°C$.

The surface that we can see is a system of cloud belts in a constant state of turmoil, with those belts nearer the equator moving slightly faster than those near the poles. Many light and dark spots appear on these cloud belts, but the most outstanding of these by far is the *Great Red Spot* which first became prominent in 1878 and has been seen pretty well all the time since then, though it varies considerably in intensity. At its most prominent it attains a startling brick-red colour. Space probe results indicate that it is a huge weather system rather like a terrestrial cyclone.

Jupiter has now fewer than sixteen moons, four of which were discovered by Galileo and are easily visible in the smallest telescope (some people claim to be able to see them with the naked eye). These four have sizes ranging from 5,150 to 3,220 km in diameter, while the other twelve are very much smaller and fainter.

Four unmanned spacecraft have visited Jupiter, each returning a great deal of new information and high quality photographs of both the planet and its moons.

Saturn

Much farther out in the depths of space lies the next giant planet, Saturn, orbiting at a distance of 1,425 million km and taking $29\frac{1}{2}$ years to travel round the Sun. Saturn also rotates on its axis in a short period – 10 hours 14 minutes – and is even more flattened at the poles than Jupiter. Saturn's polar diameter is 108,120 km while the equatorial diameter is 120,840 km. Like Jupiter, Saturn seems to be a largely gaseous body, possibly having a solid or metallic hydrogen core about 40,200 km across. Most of the ammonia seems to have frozen out of Saturn's atmosphere at the very low temperature of $-170°C$ ($-240°F$), and so the principal component gases are hydrogen and methane. A fascinating thing about Saturn is that its density is less than that of water, with the result that, although Saturn's volume is 700 times that of the Earth, its mass is only 95 times greater. The surface is once again cloud layers, but these are much less active than those on Jupiter and spots are rare.

The most interesting thing about Saturn is its curious system of *rings*. These were first seen by Galileo, but his telescope was not good enough to show their true nature, and it was left to the Dutch astronomer, Huyghens, to realize this in 1659. The ring system extends over a diameter of no less than 273,500 km from rim to rim, whereas the thickness of the system seems to be only a matter of 15 km

Seen edge-on, the rings of Saturn almost disappear.

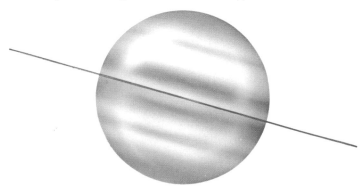

or so – the rings vanish in all but the largest telescopes when seen edge-on. The rings have several divisions in them, but can be classified into three main sections: the outer ring A, the bright ring B, and the faint inner ring C or *Crepe Ring*, the division between rings A and B being the *Cassini Division*, some 2,730 km wide. The nature of the rings is a mystery but it is thought that they may be ice particles or fragments of a disintegrated moon. The Voyager 1 & 2 space probes revealed a great deal of delicate structure in the rings.

At one point in Saturn's orbit the south face of the rings is seen from Earth, while half an orbit later the north face is seen. Thus at some times they are seen wide open, while at others they are edge-on and disappear in all but the largest telescopes because they are so narrow.

The divisions in the rings are caused by the gravitational forces of Saturn's 20 moons, one of which, Janus, was discovered in 1966. Saturn's largest moon, Titan, is 5,150 km in diameter (the biggest in the solar system) and seems to have a slight atmosphere. Titan, which was discovered by Huyghens, is easily seen with a small telescope. The other moons are fainter but several can be seen telescopically.

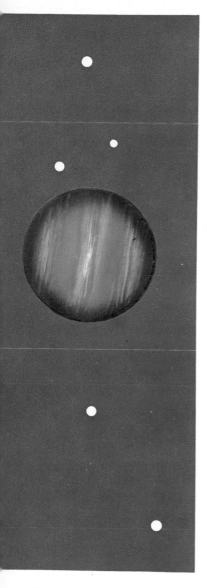

Uranus

Uranus was the first planet to be discovered telescopically. The discoverer was William Herschel. A huge planet 46,500 km in diameter, it moves so far (2,870 million km) from the Sun that it takes 84 years to complete its orbit. It is built on the same pattern as Jupiter and Saturn and is flattened at the poles to a similar extent as Jupiter. The rotational period is about 16 hours and the temperature is low, about −210°C.

A strange thing about Uranus is the tilt of its axis. Most planets have their axis of rotation nearly perpendicular to the plane of their orbit (the Earth's axis is tilted at $23\frac{1}{2}°$), but Uranus has its axis tilted at 98° so that its axis lies almost in the same plane as its orbit; thus we can sometimes see Uranus 'pole on', and at other times its equator is presented.

Uranus has fifteen moons, ten of which were discovered by the Voyager 2 spacecraft when it passed by the planet in January 1986. The largest of the Uranian

Uranus

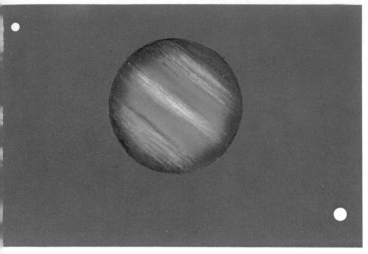

Neptune

moons is Oberon, which is about 1,550 kilometres across and the smallest, Cordelia, is only 15 kilometres across.

Neptune

Neptune is much farther out than Uranus, moving in an orbit 4,495 million km from the Sun and taking 165 years to complete it. Very similar to Uranus, Neptune has a diameter of 44,570 km but is less flattened than the former and also rather more dense, at 2·2 times the density of water. Few surface details can be seen. One of Neptune's moons is in an orbit so eccentric that it is sometimes less than a million kilometres from the planet and at other times 9 million. Neptune was discovered in 1846, but its existence was predicted on theoretical grounds in 1843 by the Cambridge mathematician John Couch Adams and the French Astronomer Le Verrier.

After its January 1986 Uranus fly-by, Voyager 2 was placed on a path that would allow it to fly past Neptune during August 1989, photographing the planet and its moons.

Pluto

The last known member of the solar system is the planet Pluto – named after the god of the Underworld – revolving around the Sun in a period of 248 years at an average distance of 5,899 million km, 40 times as far away as the Earth. Pluto follows a curious orbit which is markedly eccentric, so much so that for part of its journey round the Sun it is actually closer in than Neptune. Pluto seems to be a small frozen body about 3,000 km in diameter and covered with frozen gases. From variations in the planet's brightness it has been deduced that it revolves on its axis in about 6 days 9 hours.

The existence of Pluto was predicted by Percival Lowell in 1914. But it was not until 1930 that Pluto was finally tracked down by Clyde Tombaugh.

In 1978, Jim Christy discovered that Pluto has a moon. Named Charon, the moon appears to be about 1,200 km across – about two-fifths

Photographs taken three days apart revealed the presence of Pluto and bore out the mathematical predictions.

the size of Pluto – and moves in an orbit just 19,000 km from the planet.

Table of planetary data

Planet	Mean distance from Sun, in millions of kilometres	Sidereal period	Synodic period	Axial rotation
Mercury	58	88 days	115 days	59 days
Venus	108	224·7 ,,	584 ,,	243 ,,
Earth	149	365 ,,	—	23 h. 56 m.
Mars	228	687 ,,	780 ,,	24 h. 37 m.
Jupiter	777	11¾ years	399 ,,	9¾ h.
Saturn	1,425	29½ ,,	378 ,,	10¼ h.
Uranus	2,870	84 ,,	370 ,,	16 h
Neptune	4,495	164¾ ,,	367½ ,,	18 h
Pluto	5,900	247¾ ,,	366¾ ,,	6 d. 9 h.

Planet	Diameter in kilometres (equatorial)	Brightest magnitude	Maximum surface temp. (degrees Celcius)
Mercury	4,880	− 1·9	400°
Venus	12,104	− 4·4	475°
Earth	12,750	—	60°
Mars	6,760	− 2·8	− 20°
Jupiter	142,700	− 2·5	− 150°
Saturn	120,840	− 0·4	− 170°
Uranus	47,600	+ 5·6	− 210°
Neptune	44,570	+ 7·7	− 230°
Pluto	3,000	+ 13	− 220°

MINOR BODIES OF THE SOLAR SYSTEM

The minor planets

There is a large gap of 560 million km between the orbits of Mars and Jupiter and this was noticed as long ago as Kepler's time; Kepler himself suspected that there might be another planet in this gap. In 1772, the German astronomer Bode pointed out a curious numerical relationship between the distances of the planets, this being known as *Bode's law*. If we take the numbers 0, 3, 6, 12, 24, 48, 96, 192, and 384, where each number after 3 is double the preceding one, and then add 4 to each of them, we get 4, 7, 10, 16, 28, 52, 100, 196, 388. Now, take the Earth's distance as being 10. The table below shows the comparison between Bode's law and the actual distances of the planets.

Planet	Distance by Bode's Law	Actual distance
Mercury	4	3·9
Venus	7	7·2
Earth	10	10·0
Mars	16	15·2
?	28	—
Jupiter	52	52·0
Saturn	100	95·4
Uranus	196	191·8
Neptune	—	300·7
Pluto	388	394·6

As you can see there is remarkable agreement between the Bode's law values and the actual distances of the planets, except for a little confusion over Neptune and Pluto. But one obvious gap remains – there is no planet corresponding to 28. Whether or not Bode's law has any real significance is questionable, but it did make people think about the gap, and in 1800 six astronomers formed a group (they called themselves the 'celestial police') to make a careful search for the supposed missing planet.

They were forestalled, however, as on 1 January 1801

Piazzi of the Sicilian Observatory picked up a small planet moving at a distance of 27·7. However, this planet, Ceres, was very small – less than 800 km in diameter – and the 'celestial police' were not convinced that this was what they were looking for. They continued to search and soon picked up three more such bodies: Pallas (1802), Juno (1804) and Vesta (1807). Since then, thousands more have been observed, mostly just large lumps of rock a few kilometres across, and it is estimated that there are about 40,000 of them, mainly confined to the zone between Mars and Jupiter. Some, however, move in highly eccentric orbits which take them quite near the sun, and at least one minor planet, Hermes, has passed within a few hundred thousand kilometres of the Earth.

The origin of these bodies remains a mystery. Possibly they are the remnants of a planet which broke up, or of material which never formed into one. The orbits of a number of minor planets do intersect at one point, indicating that the former be the case. But the question remains undecided.

Most of the minor planets lie between Mars and Jupiter.

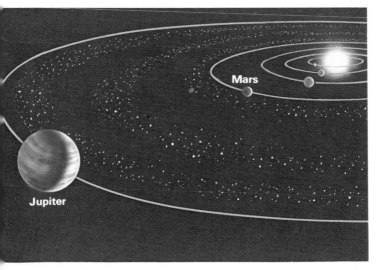

Comets

It cannot be denied that when a bright comet appears in the night sky, it receives a great deal of attention, not only from astronomers but also from anyone who cares to cast their eyes skyward. The rarity and strange appearance of comets has, over history, made them omens of good or evil. Until their nature was understood comets were regarded as signs of impending disaster. For example, a bright comet appeared in the sky in 1066 just before the death of King Harold of England. It is depicted on the famous Bayeux Tapestry.

Comets look rather like fuzzy stars with long nebulous tails that sometimes can stretch across the entire sky. Much of our knowledge of the orbits of comets can be attributed to the work of the famous English astronomer Sir Edmond Halley, a contemporary of Sir Isaac Newton. Halley calculated that many comets must move in orbits that are highly elliptical. In particular, he noted that the appearance of the

(*Above*) Brooks' Comet
(*Below*) The Great Comet

comet of 1682 was very similar to those of 1607 and 1531. He concluded that the same comet was responsible for each of these appearances, and further predicted that it would reappear in late 1758 or early 1759. Alas, Halley died some years before the comet reappeared early in 1759. Ever since then, the comet has been known as Halley's Comet. Its most recent appearance was in 1986 and it is due again in 2061, its period being about 75 years. At perihelion, the closest approach to the Sun, Halley's Comet moves inside the orbit of Venus. At aphelion, when it is farthest from the Sun, the comet moves beyond the orbit of Neptune. It is Halley's Comet that is on the Bayeux Tapestry, seen during its 1066 return.

Comets consist of a nucleus made up mainly of water ice mixed in with rock dust and various gasses. When they are far from the Sun they are virtually undetectable. But when they approach the Sun, the ice begins to melt. Steam, gas

(Top) Halley's Comet
(Centre) Orbit of Halley's Comet
(Below) De Chéseaux' Comet

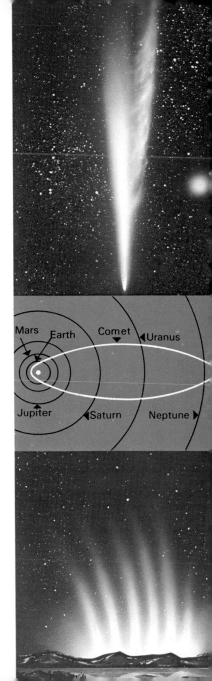

and dust are liberated and form a tenuous 'shell' around the nucleus which is known as the coma. Only when the coma begins to form can we hope to detect an approaching comet. The coma can measure, on average, 120,000 km in diameter although the nucleus itself measures only a few tens of kilometres across. Radiation pressure from the Sun pushes some of the material from the comet and the tail forms. This is why comet tails always point away from the Sun. Because of this a receding comet seems to be moving 'backwards'. Each time a comet appears it loses material, which it cannot regain. So eventually they just evaporate away and cease to exist.

During its 1986 appearance, Halley's Comet was visited by no less than five unmanned spacecraft – two from Japan, two from Russia and one from Europe. The European probe, named Giotto, passed through the coma of the comet and went as close as 500 km from the nucleus, at a speed of 76 km per second. Giotto sent back much valuable information

A meteor exploding

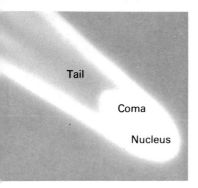

The structure of a comet

Because of perspective, meteor showers appear to originate from a single point, called the *radiant*.

including a series of spectacular pictures which showed the nucleus to be shaped rather like a giant peanut shell. Jets of material spewing from cracks in the nucleus could also be seen in the images. This is the material that forms the tail.

Meteors

Meteors or 'shooting stars' look very splendid as they flash briefly across the night sky. They are nothing whatever to do with stars, however, being simply tiny pieces of matter rushing through space and being burned up by friction as they hurtle into our atmosphere at speeds of up to 70 km per second. Most meteors are extremely tiny, being on average about the size of a grain of sand. In fact, something like 100 million of them enter our atmosphere daily. Besides lone (sporadic) meteors, there are definite showers of meteors which occur at regular intervals and are due to large numbers of these particles moving in groups in orbit round the Sun.

The Arizona Meteor Crater

Meteorites

There are some objects which are quite large, and sometimes these may reach the ground, in which case they are known as meteorites. Very often, in fact, the meteorite will explode before reaching ground level and scatter its fragments over a large area, the passage through the atmosphere of such a meteor being marked by a very bright trail and audible bangs. Meteorites are relatively rare, and since there is more sea than land on the Earth, the majority fall into the sea. However, some large ones have been found, such as the Hoba West meteorite in South Africa, which weighs 61 tonnes, and a Greenland meteorite of 37 tonnes.

Very rarely, really huge meteorites hit the earth. In 1908, a meteorite fell in Siberia the impact from which blew trees flat for 95 km around. The explosions were heard up to 900 km away, and even 160 km away men were bowled over by the shock wave. Some people think that this was the nucleus of a small comet.

The Arizona meteorite crater is one and a quarter km in diameter and 180 metres deep, and was due to a prehistoric meteorite which must have weighed about 51,000 tonnes and

74

(*Left*) The Greenland meteorite which weighs 36 tonnes
(*Right*) The etched surface of an iron meteorite

vaporized on impact. No sizeable fragments have ever been found.

The chances of being hit by a meteorite are too remote to consider, and there is no recorded case of anyone being struck directly by one. But if a meteorite of the proportions of the Arizona or Siberian ones did hit a city, the effect would be very like that of a nuclear bomb and the devastation would be colossal.

Most meteorites weigh only a few grams, and fall into one of two classes: stony meteorites, composed principally of rock, known as *aerolites*, and nickel-iron meteorites, known as *siderites*. Most of the large meteorites are in fact siderites, as the stony ones tend to break up easily. A few years ago, meteorites were found which contained organic materials, a discovery which would imply that life must form even on lumps of rock in space; but this material is now thought to have been impressed on the meteorites on landing. *Tektites*, found mainly in Moldavia and southern Australia, are small rounded pieces of a glass-like substance and are thought by many to have an extraterrestrial origin.

THE STARS

From a casual glance at the heavens it is easy to appreciate the ancient view that the stars were holes in the firmament through which shone light from beyond. However, stars are much more interesting than that. All stars are basically huge balls of gas just like the Sun, deriving their energy from the same sort of processes and radiating energy in a similar fashion. Although the basic structure of stars is similar, there are tremendous variations in size, temperature, colour, mass and many other features. Even the naked eye can distinguish considerable differences in the colour and brightness of stars.

To the naked eye, stars appear simply as points of light, and it may come as a surprise to some to learn that telescopically, too, they appear as no more than points of light. The reason is that the stars are at such immense distances from the Earth that no telescope yet built will show them as noticeable discs.

A glance at the sky suggests that there are innumerable stars visible, but in fact the average number of stars that an ordinary observer can distinguish without a telescope on a good night is only about 2,000, and the total number of stars in the entire sky which could be seen with the naked eye is between 5,000 and 6,000. The brighter naked-eye stars seem fairly evenly distributed, but with telescopic aid it becomes apparent that a great many more stars are congregated around that faint misty band of light called the Milky Way, and indeed the Milky Way itself is seen to consist of millions upon millions of faint stars.

The positions of the stars are measured on a coordinate system rather like latitude and longitude on the Earth. It is convenient to imagine that the stars are points on a huge sphere, the *celestial sphere*, with the Earth at the centre. The zero line of celestial 'latitude' – *declination* – is the celestial equator, simply a projection of the Earth's equator on to the celestial sphere, while the zero of celestial 'longitude' – *right ascension* – is the *vernal equinox*, now in the constellation Pisces, which is where the Sun crosses the celestial equator in the northern hemisphere spring.

Celestial
north pole

Celestial
equator

N.
pole

London 57°

S.
pole

16° 30′

Celestial
south pole

Sirius

Temperature: 25,000°C	11,000°C	6,000°C	4,000°C	3,000°C
Typical star: Spica	Sirius	Sun	Arcturus	Betelgeuse

The colours and temperatures of stars

The brightnesses of stars are measured on the *stellar magnitude scale*, whereby bright stars are of low numerical magnitude and faint stars of high numerical magnitude. Thus a bright star such as Aldebaran in Taurus is of 1st magnitude whereas a star just on the limit of naked eye vision is of 6th magnitude and is, in fact, one hundred times fainter – the scale has been so arranged that each unit of magnitude corresponds to a difference in brightness of 2·52 times. The relationship between magnitude and brightness,

The numbers of stars at different magnitudes

Magnitude	Number of stars
1	20
6	3,000
11	800,000
16	50,000,000
21	1,000,000,000

taking 1st magnitude as equal to brightness 1, is shown below.

Magnitude:	1	2	3	4	5	6
Brightness:	1	1/2·5	1/6·3	1/15·9	1/39·8	1/100

Stars as faint as 23rd magnitude can be observed with the 5 metre Palomar telescope, these being no less than 6,300 million times fainter than 1st magnitude stars. There are also quite a few stars brighter than 1st magnitude, and so the scale goes to 0, −1, and so on. Sirius, the brightest star, is of magnitude −1·4, whilst Venus can attain −4·4.

The distances of stars are so vast that they are measured in light years. Light travels in space at a speed of 300,000 km per second, and a light year is the distance light travels in one year (about 9·6 million million km). Even on this scale the nearest star is over 4 light years away. Another frequently used unit of distance is the parsec, which is the distance at which a star would have to be to show a parallax of 1 second of arc. One parsec equals 3.26 light years. A comparison of the apparent and absolute magnitudes of a star gives a measure of the star's distance, and the difference between these two magnitudes is often called the distance modulus.

The early astronomers took it that the fainter a star was, the farther away it was, and as a general guide this is not too bad. However, stars do vary considerably in the amount of light they radiate and so it became necessary to have some means of measuring the actual amount of radiation given off by a star. To facilitate comparison, the concept of *absolute magnitude* was introduced. The absolute magnitude of a star is the stellar magnitude it would appear to have if it were at a distance of 10 parsecs (32·6 light years) from the Earth.

Thus Aldebaran, actually of magnitude 1·1 and 21 parsecs distant, would, if placed 10 parsecs away, appear to be of magnitude 0·5, and this is its absolute magnitude. The Sun has an absolute magnitude of 4·8 and so is inherently nearly a hundred times fainter than Aldebaran.

Stars come in all colours from dull red to blue, cor-

Arcturus

Sun

Betelgeuse

61 Cygni

responding to a progression
from cool to very hot stars,
from below 3,000° to over
30,000°C. The colours are
not always easy to see without a telescope.

The sizes of stars vary
enormously; the largest stars
known can be hundreds of
times larger than the Sun.
For example, the red giant
star Antares is about 290
times larger than the Sun, so
that if the Sun and solar
system were placed at the
centre of it, Mars would still
be within the globe of the
star. The smallest stars observed are considerably
smaller than the Earth, these
being some of the white
dwarf stars.

However, the masses of
stars do not vary by nearly
so much, with the result that
stellar densities vary tremen-
dously. Betelgeuse, a huge
red star some 250 times the
Sun's size, has an average
density of less than 1/100,000
that of the Sun and so is very
rarefied indeed. On the other
hand, some white dwarf stars
have their matter com-
pressed to truly incredible
densities. Imagine the mass
of the Sun compressed into
a body far smaller than the

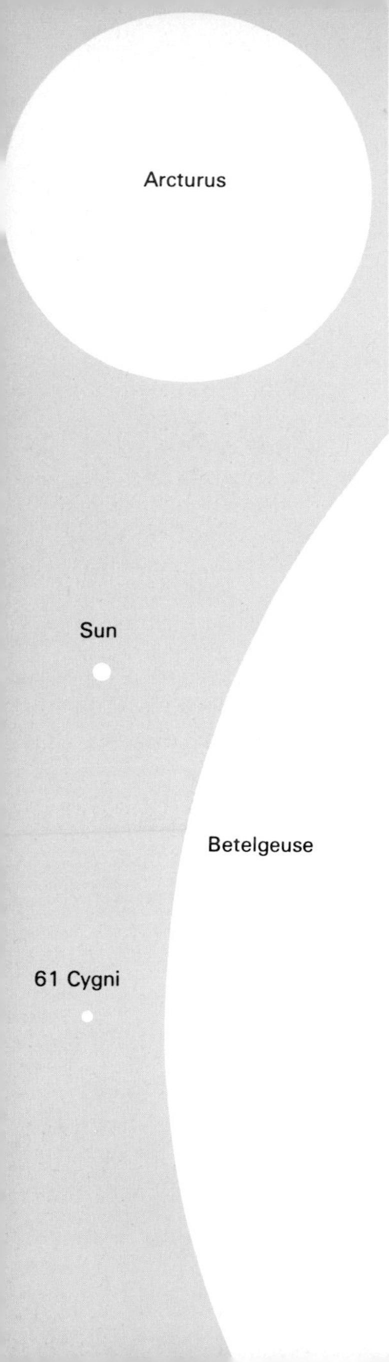

Comparative sizes of some stars

Earth – this is the case with Kuiper's Star, a white dwarf whose matter is so dense that a cubic cm of it would weigh hundreds of tonnes. The densest known stars are the pulsars, strong radio stars composed of neutrons.

All the stars are in motion with respect to each other. The Sun, for example, is moving through space in the direction of the constellation Hercules at a speed of about 19 km per second. We can regard this motion as consisting of two components – radial and transverse motion. The *transverse motion* is simply movement across our line of sight and is shown up by the small apparent motions of the stars on the celestial sphere. *Radial motion*, however, is motion directly away from us, and it is hard to see how this could be measured. The solution lies in the Doppler effect. When a star is moving away from the Earth, the Fraunhofer lines in its spectrum are displaced nearer to the red part of its spectrum, and the faster the star goes, the greater is this displacement. Thus by measuring the positions of the spectral lines, the star's radial velocity can be found.

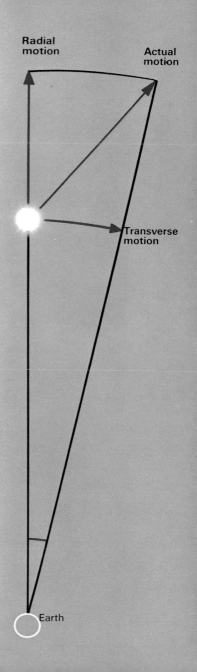

Double stars

Through the telescope there are many pairs of stars which seem to be associated with each other. Some of these pairs are simply *optical doubles*, where two stars are in fact very far apart but lie in the same line of sight and so seem close together. Many others, however, are true double stars and revolve around each other – or more properly, about their common centre of gravity. In fact, it is principally by observing the motions of double stars and the effect that each star has on the other (due to their gravitational forces) that estimates of the masses of stars have been made. Many of these double, or *binary*, stars have their components well separated and it is easy to plot their motions and orbits. However, there are many pairs which are too close together to be separated even by the most powerful telescopes, but in this case the Doppler effect comes to our rescue once again.

(Top) Position of Mizar and Alcor in Ursa Major
(Left) The wobbling motion of Sirius over the years indicates the presence of a companion star.

If two stars are revolving about each other, then at any time one of the stars will be coming towards us and the other will be moving away, while later on, the positions will be reversed. Since these two stars are so close together, then we shall see them as one. However, if we analyze the spectrum of this 'star', we shall see two sets of Fraunhofer lines, and the motions of the two stars will cause these lines to shift about with respect to each other; from this we deduce that we have a binary pair.

In some cases the presence of an invisible binary companion to a known star can be detected by the wobbling motion of the star through space.

The best known double star visible to the naked eye is the pair Mizar and Alcor in the constellation of the Great Bear. Through a telescope another star can be seen, as Mizar itself is a double; the situation is even more complicated, as the bright component of Mizar is a spectroscopic binary too, and so the whole system is made up of four stars. Multiple stars of this type are quite common and present a beautiful sight.

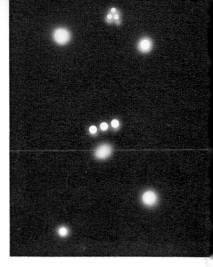

The position of the multiple star θ Orionis, nicknamed the Trapezium because of the arrangement of its four main components. It is surrounded by a gaseous nebula.

The 'double-double' in Lyra

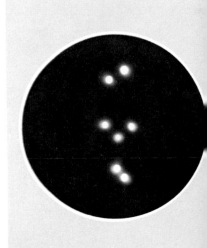

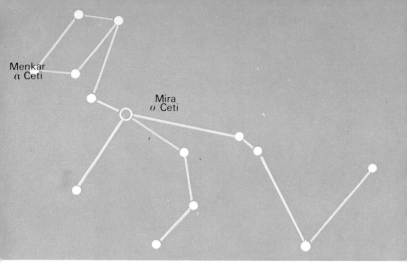

The position of Mira Ceti

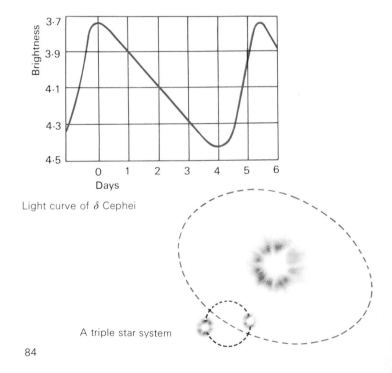

Light curve of δ Cephei

A triple star system

Variable stars

By no means all stars are as constant in their light output as the Sun, and a great many vary quite considerably, some regularly and others irregularly. One of the best known variable stars is Mira ('the wonderful') in the constellation Cetus. The curious behaviour of this star was noted hundreds of years ago. Although it has been known to attain 3rd magnitude, it can slip down to magnitude $9\frac{1}{2}$ (a difference in brightness of 300) and tends to be invisible to the naked eye, for most of its period of fluctuation, which is about 47 weeks. Mira Ceti, incidentally, is a red giant star. There are many 'long-period variables' similar to Mira.

Another type of variable altogether is the *irregular variable*. Such stars have no regular period at all and astronomers just do not know what they may do next. Betelgeuse, the red giant in Orion, behaves in this fashion, although the variation is not great. Amateurs can contribute a lot to the observation of irregular variables.

Other variables have completely predictable periods. One of this type is the *eclipsing binary*. If two close stars revolve around each other and the plane of their orbits is correctly placed, one will pass in front of the other at certain times and so reduce the total amount of light coming our way, and the binary will appear to be a variable star. Algol ('eye of the demon') in Perseus is a fine example with a period of just under three days. There are also some cases of eclipsing binaries in which one of the component stars is itself intrinsically variable.

Of great importance to the astronomer are the *Cepheid variables*, so named after the first of their type observed, δ Cephei. These show very regular variations, climbing sharply to a maximum and dropping off slowly, with periods ranging from a day to several weeks. Cepheid variables are in fact pulsating stars – they expand and contract, giving off the most light just before they are at their largest. There are subgroups of these, such as the RR Lyrae stars whose periods are only a matter of hours, and the W Virginis variables (each type being named after the first of its class to be observed).

The importance of the Cepheid variables lies in the

definite relationship between their periods and their absolute brightnesses. Thus, by comparing the absolute magnitude of a Cepheid (found from its period) with the observed magnitude, its distance can be found directly. Measuring stellar distance is a tricky problem as the method of parallax can be used only for the nearest stars. Farther out than about 200 light years the method becomes decidedly inaccurate and statistical techniques have to be used. The *period-luminosity law* of the Cepheids is thus a valuable tool for measuring not only the distance of the Cepheid itself, but also of the star cluster or galaxy in which it may be situated. Likewise, the RR Lyrae variables can be employed.

There are still other types of variables, the catastrophic variables – *flare stars*, *novae*, and *supernovae*. Flare stars are usually dim red dwarfs, very faint stars which can some-

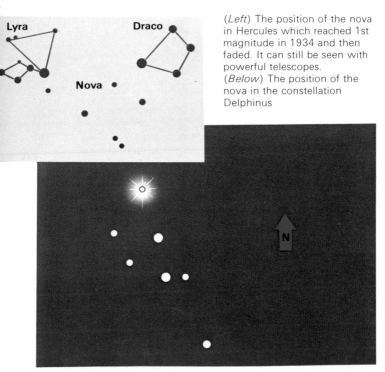

(*Left*) The position of the nova in Hercules which reached 1st magnitude in 1934 and then faded. It can still be seen with powerful telescopes.
(*Below*) The position of the nova in the constellation Delphinus

The Crab nebula

times brighten by several magnitudes in a matter of minutes. Some outburst on the star's surface is responsible, and indeed our own Sun exhibits flares, on a very much smaller scale.

Novae are stars which blaze up suddenly, taking only a few days or less to reach maximum brightness, and then fade away slowly into obscurity. Because it is impossible to predict when and where novae may appear, it is often amateur observers who discover them. For example, the English amateur George Alcock discovered a naked-eye nova in the constellation Delphinus as recently as July 1967. Novae are thought to be stars which, due to some instability, blow off an outer shell of gas, increasing their brightness by 20,000 times in the process, and then settle down to a stable state again. Many novae are thought to be recurrent. Supernovae, which may brighten by a factor of hundreds of millions (some supernovae in other galaxies have briefly outshone their galaxies), are true celestial disasters and are due to the almost complete disintegration of stars in awe-inspiring explosions.

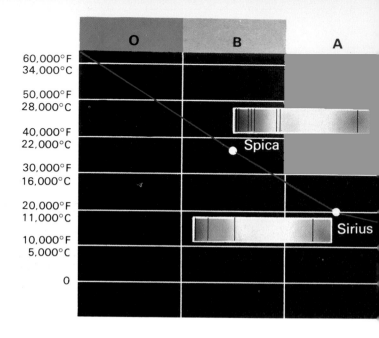

Stellar spectra

Nearly everything we know about stars has come from analyzing their spectra. Stars emit radiation over a wide range of wavelengths but do not radiate equal amounts at each wavelength. Each star has a characteristic wavelength at which it emits a maximum amount of energy, and the more energetically a star is radiating, the nearer to the blue end of the spectrum will this maximum occur. This is why hot stars appear blue and cool stars appear red. The Sun is quite average, being yellow in hue.

Associated with this temperature progression are varying sets of spectral lines. For example, in very hot stars hydrogen lines are prominent, together with other ionized gases (gas atoms are made up of a nucleus surrounded by electrons – an atom which has lost one or more electrons is 'ionized'), while in stars like the Sun metals become common, and in cool red stars lines due to elementary molecules such as titanium oxide begin to appear. In very hot stars molecules

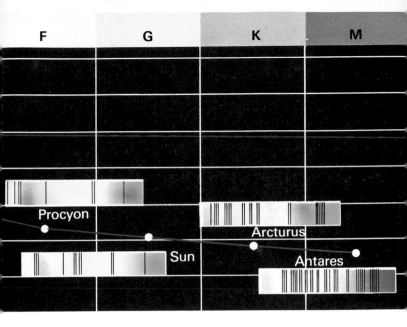

Several spectra and magnitudes of stars

(orderly combinations of atoms) cannot survive.

Several attempts have been made to classify stars into convenient groups, the system now adopted being based on a scheme evolved in 1890 by Pickering at Harvard. From the hottest to the coolest, stars are grouped into one of the following classes: W, O, B, A, F, G, K, M, R, N, S. Only 1 per cent of stars fall into classes W, R, N, S combined. A useful way of remembering this classification is to recall the mnemonic 'Wow! Oh Be A Fine Girl Kiss Me Right Now Sweetie', or something like that. The classes are each split into ten subgroups, numbered 0 to 9. The Sun is of spectral class G2, being very similar to the bright yellow star Capella (G0), while Sirius is of class A1 and Betelgeuse is of class M2. The exceptionally hot W stars have emission line spectra. Odd stars are designated p (for 'peculiar') and dwarfs, giants and supergiants are denoted by d, g and s, respectively, placed before the main letter.

The Hertzsprung-Russell diagram. This shows the relationship
between the spectral class and absolute brightness of stars. The top
diagram shows position of stars on the graph schematically; the
bottom diagram shows the actual position of some stars.

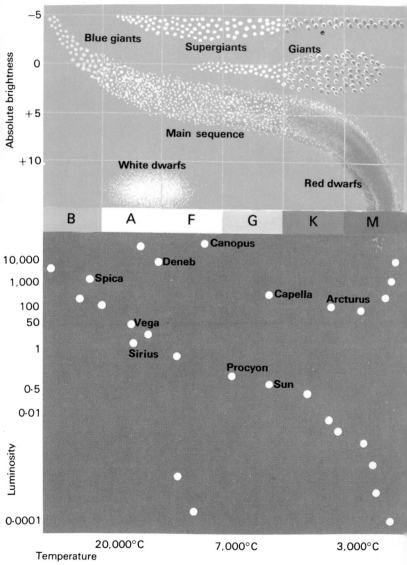

Colour-magnitude and the evolution of stars

A valuable aid to the study of stellar types and evolution is the colour-magnitude diagram, known as the *Hertzsprung–Russell* diagram. This is a plot of the absolute brightness or luminosity of stars against their spectral class, colour or temperature. Most stars lie on a band running from the top left (hot, bright) corner to the bottom right (cool, dim) corner, known as the *main sequence*. Giants and supergiants lie above the main sequence and white dwarfs below it. Incidentally, the Cepheid variables too lie above the main sequence. The Sun sits very comfortably near the middle.

It is believed that stars form from clouds of gas and dust. As the cloud breaks up into fragments, these fragments begin to contract under gravitational forces, heat up and radiate heat. As the temperature builds up sufficiently for nuclear reactions to begin, the star moves towards the main sequence and settles down to a stable state at some position on it, depending on its mass and temperature. It remains in this position for the major part of its life, generating energy by conversion of hydrogen to helium (as the Sun does), but when it has used up most of its hydrogen fuel, it moves off the main sequence and expands into a red giant. After this, what happens is not well understood, but it is thought that the star rapidly goes through different fuels and generates all sorts of heavy elements until it becomes highly unstable and explodes, as a nova or a supernova.

The life cycle of a star

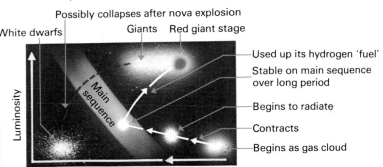

Possibly collapses after nova explosion

White dwarfs   Giants   Red giant stage

Main sequence

Luminosity

Temperature

Used up its hydrogen 'fuel'

Stable on main sequence over long period

Begins to radiate

Contracts

Begins as gas cloud

OUR GALAXY

The galaxy of which our Sun is a member consists of much more than just individual stars, of which there are about a hundred thousand million. Many stars are gathered together into groups or clusters containing anything from a few tens of stars to a hundred thousand or more.

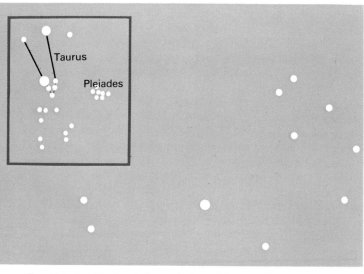

The Pleiades. The inset shows their position near Taurus.

The globular cluster M13 in Hercules

Star clusters

One type of cluster is the *open cluster*, where the stars are not very densely packed together and are only loosely associated. The best-known open cluster is the *Pleiades*, or Seven Sisters, clearly visible to the naked eye not far from the bright red star Aldebaran in the constellation Taurus. They were named after the seven daughters of the mythological Atlas and Pleione. Only six of the stars are clearly visible – whether this has anything to do with the misbehaviour of the seventh daughter is open to question. Telescopic observations show several hundred stars in this group.

At least 350 open clusters are known, containing anything from a few to several hundred stars. It appears that the member stars were all formed in the same region at about the same time and have been loosely held together by gravitational forces. Such clusters tend to break up with time as the individual stars overcome the weak pull of the others.

Another type of cluster is the *globular cluster*, of which about a hundred are known, each containing something like a hundred thousand stars very closely packed together in comparison with the stars in the vicinity of our Sun. Globular clusters are impressive sights when seen through a telescope. The best-known examples are in the constellations Hercules and Canes Venatici. Globular clusters contain many red stars, and also many RR Lyrae variables which are useful for determining the clusters' distances.

The Milky Way

The Coal Sack in Crux

The Pleiades, showing nebulosity

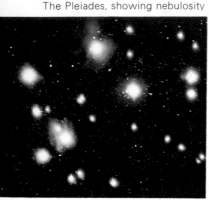
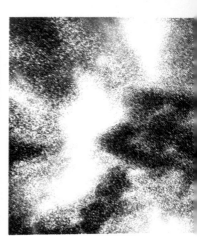

The Great Nebula in Orion

Nebulae

Looking at the Milky Way, which contains most of the stars in our galaxy, we see many hazy light patches and irregular dark patches. These are the nebulae (from the Latin for 'clouds'). The French astronomer Messier, who was interested in comets, found these objects a nuisance and set about cataloguing them so as to avoid confusion with comets. Messier's catalogue showed 103 of these objects, some of which are other galaxies beyond our own. Since then, many more have been found.

The true nebulae are composed of gas and dust – the stuff from which stars form – and can be split into two classes, *emission* and *reflection nebulae*. Emission nebulae shine because they absorb energy from nearby stars and re-emit this as visible light, giving rise to an emission spectrum in the process from which their composition can be deduced. The gas of which nebulae are composed is principally hydrogen.

Reflection nebulae, on the other hand, shine by light from nearby stars which is reflected from the dust particles in the clouds. There is no difference in composition between emission and reflection nebulae – it is only when the nebula is close to very hot stars of O and B type that the absorption and re-emission process works.

A really spectacular example of an emission nebula is the Great Nebula, M42, in Orion. This is just visible to the naked eye as a faint hazy patch in the 'sword' of Orion, but it is seen through the telescope to be a huge glowing mass of gas with many hot stars embedded in it. In fact, this cloud of gas is about 26 light years across and about 1,600 light years distant.

Reflection nebulae are not so easily seen, but on photographs the stars of the Pleiades, for example, are seen to be surrounded by nebulosity of this sort.

Many *dark nebulae* can be seen in the Milky Way without a telescope, for example the Coal Sack which appears as a black area in the Milky Way close to the Southern Cross. Dark nebulae are the result of dust particles obscuring the light from the stars beyond; they are not the result of a lack of stars.

Interstellar dust and gas

Dark nebulae are areas of gas and dust of relatively high density, although still exceedingly low in density compared with our air (the density of these clouds is $1/10^{24}$ that of water). But most of the galaxy seems to have dust spread very thinly through it, and this has a noticeable effect on starlight. The effect of this dust is to absorb some of this starlight and to scatter some of it away from our direction; and so distant stars seem to be fainter than they really ought to be. This has led to some errors in the determination of distances. Another effect is to make stars appear more red than they really are. The dust scatters short-wave light such as blue much more than it scatters long-wave light such as red. The result is that the spectra of stars have the red part

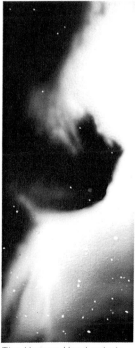

The Horse s Head nebula

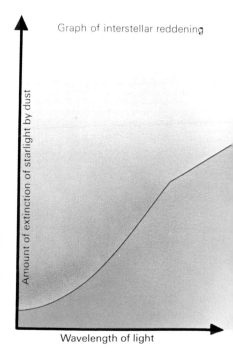

Graph of interstellar reddening

Amount of extinction of starlight by dust

Wavelength of light

M16, the nebula in Scutum

accentuated slightly, all of which makes spectral analysis harder.

The dust particles seem to be made of carbon in the form of graphite; they may be blown out into space from the atmospheres of N-type stars. The particles are tiny, smaller than the wavelength of visible light; but, spread over vast volumes of space, their effect is quite pronounced. The effect of this dust is that the centre of our galaxy is totally invisible to optical telescopes.

However, the interstellar hydrogen gas emits radio waves of 21-centimetre wavelength and these are not at all affected by the dust, being more than a hundred thousand times longer than visible light. By using radio telescopes, it has been possible to map the distribution of hydrogen gas in our galaxy and thus to find out much more about its shape. This gas, incidentally, makes up about 10 per cent of the total mass of our galaxy, and the dust only about 0·01 to 0·1 per cent, which emphasizes how efficiently it obscures light.

There is probably as much matter in diffuse form in interstellar space as there is condensed into stars. It is only in the past few decades that astronomers have reached this conclusion, and it has great significance in discussions of the formation and age of the universe.

Our galaxy, edge-on and (*opposite page*) from 'above'.

The structure of our galaxy

In general shape, our galaxy bears a close resemblance to a fried egg. The centre of the galaxy is a flattened sphere of stars, and most of the other stars are confined to a thin disc around this (the white of the egg, if you like). The total diameter of our galaxy is about 100,000 light years, more usually referred to as 30,000 parsecs, and the Sun lies about three-fifths of the way from the centre to the edge at a distance of some 9,000 parsecs from the centre. The thickness of the galactic disc in our locality is only about 400 parsecs. The central nucleus of the galaxy is about 5,000 parsecs thick and lies in the direction of the constellation Sagittarius.

The whole galaxy is rotating in a period of about 220 million years, and so the Sun, and the Earth with it, is rushing around the galactic centre at a speed of about 290 km per second. Incidentally, nearly four-fifths of the mass of

The arrows indicate the position of the solar system.

the galaxy is concentrated in the central part. Above and below the plane of the galaxy lie the hundred or so globular clusters spread out in a huge halo.

The matter and stars in the galactic disc are not evenly distributed but grouped largely into 'arms' of material which radiate in spiral shape from the centre. Our galaxy is, in fact, a spiral galaxy like the one described in the next chapter. The Sun itself seems to lie just on the edge of one of these spiral arms.

The galactic nucleus consists largely of old red stars (known as Population II stars) like those in globular clusters, and very little gas and dust.

The stars in the spiral arms tend to be rather younger with more blue stars (the stars in the arms are Population I), and nearly all the gas and dust is concentrated here. The process of stellar formation is still taking place quite rapidly in the arms.

GALAXIES AND THE UNIVERSE

In the eighteenth century Herschel suggested that some of the nebulae might be distant 'island universes' like our galaxy, but it was not until the advent of the 2·5 metre reflector on Mount Wilson that Hubble was able in 1923 to measure the distance of one of these objects and show that it was well beyond our own galaxy.

The galaxy whose distance he measured was the Great Nebula in Andromeda, which is just visible to the naked eye as a faint patch in the sky. Hubble managed to identify Cepheid variables in the Andromeda galaxy, and from their periods he deduced that this galaxy must be about 750,000 light years away. Thus the Andromeda galaxy seemed to be somewhat smaller than our own. However, in 1952 it was found that the Cepheid variable scale was in error – Cepheids of Population I were found to be brighter than those of Population II, and Hubble had assumed that the Cepheids he had measured were of the Population II variety. Thus the distance of the Andromeda galaxy was revised to

The Andromeda galaxy

over two million light years and astronomers realized that it was rather larger than our own galaxy. Obviously even a body as large as our galaxy has no particular importance in the universe!

The Andromeda galaxy is very similar to our own in many respects. Like ours it is a spiral, and like ours it has two smaller satellite galaxies. Our satellite galaxies are the Large Magellanic Cloud and the Small Magellanic Cloud, easily seen from the southern hemisphere.

Galaxies come in many shapes and sizes. Many are *spiral galaxies* like our own, though identification is sometimes difficult as we see them from all sorts of angles, and it is hard to tell if a galaxy seen edge-on is spiral. Besides spirals there are *irregular galaxies*, such as the Magellanic Clouds, which have no perceptible ordered structure. *Elliptical galaxies* are quite common, looking not unlike the globular clusters, and there are also *barred spirals* whose spiral arms

(*Above*) Elliptical and irregular galaxies
(*Below*) Elliptical, spiral and barred spiral galaxies

Possible colliding galaxies

A group of galaxies in space

spring from the ends of a central bar of material instead of from an ellipsoidal nucleus. All galaxies contain thousands of millions of stars and varying quantities of other matter such as gas and dust.

More and more galaxies become visible right out to the limits of detection as bigger and bigger telescopes are used, and it is estimated that about a thousand million can be seen with the 5 metre Palomar Mountain reflector. The distances involved are staggering. It is estimated that galaxies up to 5,000 million light years away can be detected at present – this means that the light which we are seeing from these immensely distant objects started on its journey towards us before the Earth was formed. Most current estimates put the age of the Earth at approximately 4,500 million years.

Measuring such distances poses great problems. It is possible to measure the distances of nearer galaxies by the brightness of their Cepheid variables, but beyond this, things are tricky.

One technique is based on the view that, by and large, the brightest supergiant stars in all galaxies have about the same intensity, and so by measuring the brightness of the brightest detectable stars in galaxies it is possible to make reasonable estimates of distances. Other statistical methods have been tried, for example using novae; but beyond 25 million light years we have to estimate distances by using the hypothesis that the total luminosity of all galaxies is about the same – they do not seem to vary by more than a couple of magnitudes.

Like stars, galaxies tend to congregate into clusters. Our own galaxy is a member of a cluster with more than twenty members (including the Andromeda galaxy), while some clusters have several hundred member galaxies. Unlike stars, however, galaxies are not, on average, very far apart in relation to their diameters. The Andromeda galaxy is less than twenty galactic diameters away, and others are closer, while in some clusters only a few galactic diameters separate the cluster members. Thus many galaxies have noticeable gravitational effects on each other, distorting the spiral arms for example, and collisions can occur.

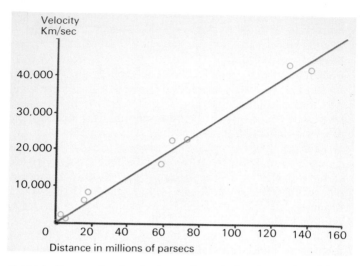

Hubble's diagram showing the increasing velocity of recession of the galaxies with distance

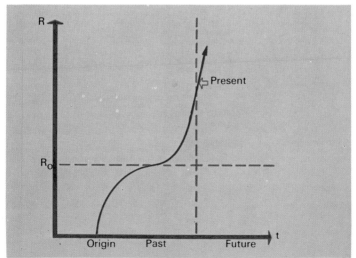

The increase in radius of the universe according to one version of the big-bang theory

The red shift

In 1920 Slipher of the Lowell Observatory in Arizona found that all the galaxies, apart from those in the local group, showed strong Doppler shifts in their spectra and must be rushing away from us at high speeds. Hubble investigated further and came to the conclusion that there was a definite rule that the farther away a galaxy was, the faster it was receding from us. Now we can observe galaxies which are receding faster than 144,800 km per second – half the speed of light. The factor relating the speed of a galaxy to its distance is known as *Hubble's constant*, and knowing this, if we now look at matters in reverse, we see that at some time in the past all the galaxies must have been packed close together. Hubble's constant can be used to give us that time, which is more than 10,000 million years ago.

This suggests that the universe may have been formed some 10,000 million years ago in an explosive fashion and has been expanding ever since. At first glance it looks as though our galaxy lies at the centre of this great expansion, and perhaps we are in a privileged position after all. But this is not so: every group of galaxies is moving away from every other group, and so *whichever group you observed from*, it would seem as though all the others were racing away from you in particular.

Various people have questioned the validity of the red shift, and it has been suggested that some agency other than velocity of recession may account for the Doppler shifts in the spectra of galaxies. It is known, for example, that a strong gravitational field will produce a red shift, but so far no satisfactory alternative hypothesis has been advanced.

The limit of detection of normal galaxies with the 5·1 metre telescope is about 5,000 million light years, but the Hubble Space Telescope will reach out much farther than this. However, there is a much more definite limitation to how far out into space we can observe: if Hubble's law of expansion continues to hold, then galaxies at a distance of more than 10,000 million light years will recede at about 300,000 km per second – exactly the speed of light – and no detectable light signal will ever reach us, no matter how large a telescope we may use.

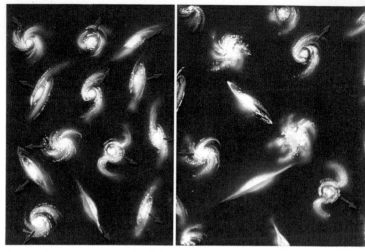

The steady-state universe. (*Left*) The galaxies are moving apart and (*right*) new matter is formed in the space between them.

Theories of the universe

How did the universe begin, and when? These questions have puzzled men for countless years. Present-day theories fall into two classes – *evolutionary* and *steady-state*.

Evolutionary theories. The Belgian George Lemaître put forward the idea that about 20,000 million years ago all the matter in the universe – enough, he estimated, to make up a hundred thousand million galaxies – was all concentrated in one small mass which he called the 'primeval atom'; this would have been of incredible density. This primeval atom exploded for some reason, sending its matter out in all directions, and as the expansion slowed down, a steady state resulted, at which time the galaxies formed. Something then upset the balance and the universe started expanding again, and this is the state in which the universe is now. There are variations on this theory: it may be that there was no steady state. However, basically, evolutionary theories take it that the universe was formed in one place at one point in time and has been expanding ever since.

Will the universe continue to expand? It may be that the

universe will continue to expand forever, but some astronomers believe that the expansion will slow down and finally stop. Thereafter the universe will start to contract until all the matter in it is once again concentrated at one point. Possibly the universe may oscillate forever in this fashion, expanding to its maximum radius and then contracting, over and over again.

The steady-state theory. Developed at Cambridge by Hoyle, Gold and Bondi, the steady-state theory maintains that the universe as a whole has always looked the same and always will. As the galaxies expand away from each other, new material is formed in some way between the galaxies and makes up new galaxies to replace those which have receded. Thus the general distribution of galaxies remains the same. How matter could be formed in this way is hard to see, but no harder than seeing why it should all form in one place at one time.

Graphical representation of the pulsating universe

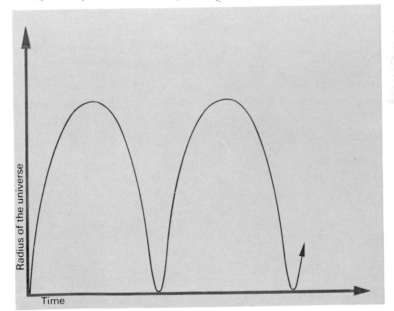

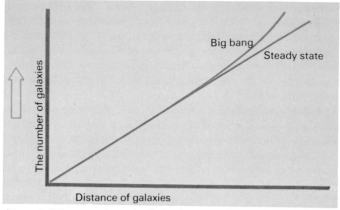

At great distance — about the limit of present telescopes — there should seem to be more galaxies in an area of sky on the big-bang theory than on the steady-state theory. In other words the galaxies — seen as they were thousands of millions of years ago — should appear closer together.

Deciding between theories

How can we decide which of these theories is closest to the truth? The method is in principle quite simple. Since the very distant galaxies are thousands of millions of light years away, then we are seeing them as they were thousands of millions of years ago. If the evolutionary theory is correct, then the galaxies were closer together in the past than they are now and so distant galaxies ought to appear to be closer together than nearer ones. According to the steady-state hypothesis there should be no difference.

Unfortunately optical telescopes cannot reach out far enough to enable us to decide. However, some galaxies emit great quantities of radio waves and can be detected even farther away than the optical range. There are still many doubts about radio galaxies, but if they are uniformly distributed, then they too should show up any difference between past density and present density. In fact, the evidence seems to suggest that there is a difference, that the galaxies were closer together than they are now, and so the evolutionary theory is partially confirmed and the steady-state theory – in its original form at least – must be rejected.

Quasars

An exciting recent development was the discovery of quasi-stellar radio sources, 'quasars' for short. In 1963 M. Schmidt at Palomar investigated the spectrum of a faint blue star which had been identified as coinciding with a radio source. He was amazed to find that this object showed a truly colossal red shift, and if it obeyed Hubble's rule, then it must be thousands of millions of light years away. This was obviously no ordinary star. Quite a few others have been found, some apparently over 10,000 million light years away, travelling at over 240,000 km per second. To be as bright optically as they are they must be 200 times as luminous as ordinary galaxies. Furthermore, they seem to be very small, a light year or so across. No one as yet can account for the production of such enormous amounts of energy. It was thought that quasars might be used to test theories of the universe, but doubts have been cast on their distances and the energies involved. It may be that the red shift in their spectra is not due to recession; if this is so, perhaps some other mechanism can also explain the red shift in normal galaxies. Several people have cast doubts on the reality of the expansion of the universe, but no plausible alternative explanation exists.

The Quasar 3C 273

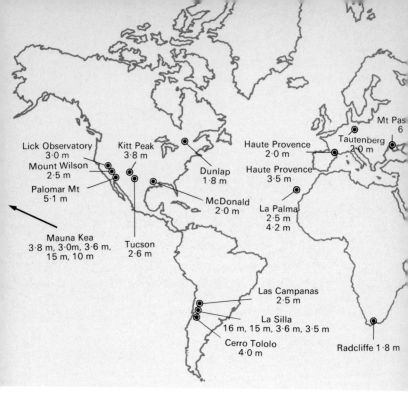

Lick Observatory
3·0 m
Mount Wilson
2·5 m
Palomar Mt
5·1 m

Kitt Peak
3·8 m

Dunlap
1·8 m

Haute Provence
2·0 m

Tautenberg
2·0 m

Mt Pas
6

Haute Provence
3·5 m

McDonald
2·0 m

La Palma
2·5 m
4·2 m

Mauna Kea
3·8 m, 3·0m, 3·6 m,
15 m, 10 m

Tucson
2·6 m

Las Campanas
2·5 m

La Silla
16 m, 15 m, 3·6 m, 3·5 m

Cerro Tololo
4·0 m

Radcliffe 1·8 m

THE FUTURE

What does the future of astronomy hold in store? It looks like ground-based astronomy will undergo another major revolution as new large telescopes come into operation. For instance, the observatory sites at La Palma (in the Canary Islands), Mauna Kea (on Hawaii) and those in Chile, all have new telescopes planned that will surpass their predecessors in terms of light gathering capability.

On Mauna Kea, which is the world highest observatory, at 4,000 metres above sea-level, the 'Keck', one of the newest telescopes, has a mirror that is ten metres across that is made up of 36 hexagonal elements. Computers constantly adjust the relative positions of the elements in order to preserve the mirror's figure accurately.

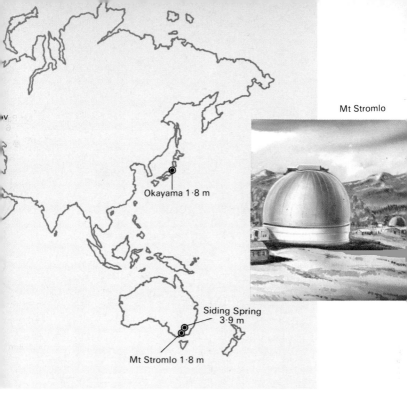

Mt Stromlo

Okayama 1·8 m

Siding Spring
3·9 m

Mt Stromlo 1·8 m

The distribution of major optical, infrared and sub-millimetre reflecting telescopes

The European Southern Observatory, at La Silla, in Chile, will soon have a sixteen metre telescope in operation. This instrument actually consists of four 8 metre telescopes acting together as one. They will also be capable of being used independently.

Far better though, would be a telescope in Earth orbit where the atmosphere could not hinder the observations. The Hubble Space Telescope (HST) is such an instrument. At its operational altitude of 500 kilometres, HST can provide images of objects seven times further away than any current optical telescope is capable of providing. Seven times further means that there will be 350 times more volume of space to observe.

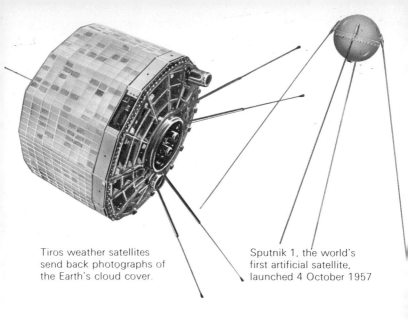

Tiros weather satellites send back photographs of the Earth's cloud cover.

Sputnik 1, the world's first artificial satellite, launched 4 October 1957

Space exploration

The Earth's atmosphere imposes limitations on what we can observe from the ground; observatories in space enable astronomers to observe more clearly and in light of different wavelengths.

The year 1957 saw the launching of the first-ever artificial satellite, Sputnik 1, by the Russians. This took the world completely by surprise and the Americans followed suit within a few months, in 1958. Before this, however, rockets had been used for astronomical observations, but they could only get results for a few minutes at the top of their flight path before they fell back to Earth again. Orbiting satellites offer a much more satisfactory means of observing astronomical phenomena, and indeed, of observing the Earth and such phenomena as weather patterns. Since the early days, several astronomical satellites have been put into orbit, and the observation that these satellites have permitted have revolutionised astronomy. Objects have been discovered that were never dreamed of previously, and new light has been shed on known phenomena that has enabled us to

Rocket and satellite

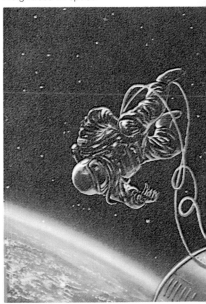
An astronaut floating weightless in space

better understand the workings of the universe. The advent of astronomical satellites has literally opened up the whole spectrum for examination. The combination of ground-based and satellite observatories means that we can examine all kinds of astronomical objects at all wavelengths from low frequency radio waves to the highest energy gamma rays.

Access to space has become much easier and more frequent, some say even routine. Heavy lift launch vehicles and the space shuttle can carry large satellites and laboratories to Earth orbit. The shuttle is even capable of returning such payloads to Earth for refurbishment and reuse.

Manned space stations look set to be the next major staging post in orbit. The Soviet Mir and Americas Freedom space stations will be places on which much research will be carried out. Later on, it will be possible to construct spacecraft designed for the planets on, and launch them from, such space stations.

The surface of the Moon

Deep space probes

In 1967 both the Russians and the Americans landed space probes safely on the surface of the Moon, taking thousands of pictures and samples and sending the results back by television and radio, and, of course, the Americans achieved manned landings in 1969. The Russians have managed to land five spacecraft on the planet Venus, while the Americans have achieved three close passes with the Mariner. Furthermore, by early 1972, four American and two Russian craft had taken close-up photographs of Mars, and in 1976 came two Viking landings. Probes have also flown past Mercury, Jupiter, Saturn, Uranus and Neptune. In addition, two comets have been visited by no less than six spacecraft, five to Halley's Comet in 1986 and one to comet Giacobinni-Zinner in September 1985. Some of these comet probes actually flew through the comets to gain detailed data.

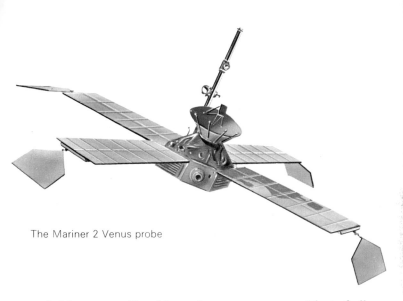

The Mariner 2 Venus probe

Achievements like this are by no means easy. First of all there are two critical speeds to consider. Firstly, from Newton's law of gravitation we find that the speed a satellite has to attain in order to stay up in its orbit is about 28,150 km per hour. Having attained this speed in the right direction, it will continue in its orbit forever, unless slowed down by thin traces of atmosphere, in which case it will eventually spiral in closer to the Earth to be burned up by friction like a meteor. Secondly, the speed a rocket must attain to escape completely from the Earth's gravitational pull – the escape velocity – is 40,200 km per hour. Even with this huge initial velocity, the rocket will slow down to crawling pace eventually and so slightly higher speeds are needed to get anywhere in a reasonable time. The problems of guiding rockets after launching are enormous and the whole business is extremely costly. Present-day rockets, using chemical fuels, can only operate for a few minutes and so must be made big enough to accelerate past the critical speed in this time, after which they coast on through space. No rocket can carry enough fuel for fully powered flights to the Moon or planets, so journeys take a long time.

Manned exploration

Probably the most spectacular manned space project to date was the Apollo lunar landing programme. Between July 1969 and December 1972, no less than twelve men had worked on the surface of the Moon. In addition to gathering samples of lunar material, they set up a series of scientific experiments which relayed back to Earth information which would allow astronomers and planetary scientists to form an accurate idea of just what the Moon is like inside. The Apollo missions also proved that it is possible to send human crews to an alien environment, have them carry out detailed scientific work, and return home unharmed.

Manned flights to Mars are now being seriously considered by both the Americans and the Russians. But they are unlikely to happen until early in the next century due to the complexity and cost of the operations involved. It may well be the case that any manned Mars mission will be carried out on an international collaborative basis in order to minimise the cost to any one nation. A trip to Mars and back will take over two years. Having arrived at Mars, a trip that would have taken several months, the crew would need to wait for Earth to be in a suitable position before they could

The Apollo spacecraft in lunar orbit.

A possible lunar base, having an underground nuclear power station with radiator to get rid of excess heat, manned quarters, and radio 'dish' (in the background) to communicate with Earth.

even begin a return flight. So there will be ample opportunities for performing a wide range of investigations of the planet over a period of weeks, if not months.

Manned missions to other planets in the solar system are another matter. The nearest planet to Earth is Venus. But the environment there is, as we learned earlier, totally hostile and it is unlikely that humans will ever land there. Exploration of that planet will be left to unmanned probes. Mercury also is out of the question. Jupiter, Saturn, Uranus and Neptune have no solid surface to land on. That leaves their moons. Manned landings on say the larger moons of these planets may be possible at some time in the distant future, but not in our lifetimes.

The next most likely phase of manned space exploration, apart from Earth orbiting space stations, will be the setting up of a permanent lunar base. It is probable that a good part of it will be buried under the surface for protection from temperature extremes and solar radiation.

THE CONSTELLATIONS

Even a casual observation of the night sky reveals that the stars appear to fall into well-defined patterns and groups, and for thousands of years men have divided the sky up into such groups, the constellations. The ancients named the constellations after mythological characters, gods and demigods, as well as everyday objects and creatures, and though the forms of some of the groups are readily apparent, a fairly colourful imagination is needed to make out many of the others. Most of the major present-day constellations were named by the Greeks, although the Latinized forms of the names are normally employed. In fact Ptolemy catalogued 48 constellations to which others have added, there now being a total of 88.

Because of the rotation of the Earth, the whole celestial sphere appears to rotate once a day, and so the constellations move across the sky. Furthermore, as the Earth moves round the Sun in its orbit, different constellations become visible in the night sky. In fact the constellations seem to rise four minutes earlier each day until, after a full year, they have returned to their original positions.

Because the Earth is a sphere, not all the constellations which are visible from one point on the Earth's surface can be seen from another point, and vice versa. At the north pole, for example, the stars follow paths parallel to the horizon (the celestial pole is directly overhead and the celestial equator runs parallel to the horizon), and these are constantly visible when the Sun has set. At the equator, however, the celestial equator runs directly overhead and the celestial poles lie on the horizon. Thus all the constellations can be seen at some time, but none are visible all the time. Stars which can always be seen are said to be *circumpolar*; at the poles, all the visible constellations are circumpolar, while none is so at the equator. In between these latitudes some of the constellations will be circumpolar, some will not. To find out which stars or constellations are circumpolar in your own latitude, simply subtract your own latitude from 90 degrees. Constellations with declination greater than your answer will be circumpolar. For example,

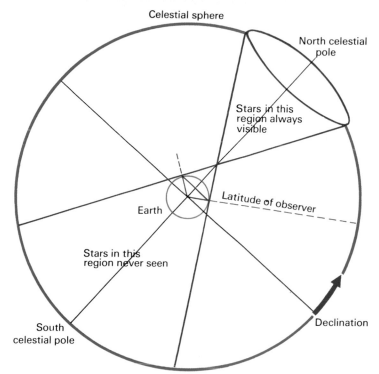

Celestial sphere

North celestial pole

Stars in this region always visible

Latitude of observer

Earth

Stars in this region never seen

South celestial pole

Declination

An observer on the Earth can see stars whose declination is 90 degrees north or south of his latitude on the Earth.

if you were at latitude 50 degrees north, then all stars with a declination of 40 degrees or more would be circumpolar. Furthermore, you would *never* see stars lying more than 40 degrees *south* of the celestial equator.

The star maps

The easiest way to learn the constellations is to begin by finding two or three immediately recognizable constellations and to use these as signposts to the other conspicuous ones. The faint, ill-defined constellations can then be filled in with ease. The first two of the star maps which follow, I have called key maps 1 and 2, and these show the constellations the Great Bear (which includes the familiar Plough) and Orion respectively, both of which groups are conspi-

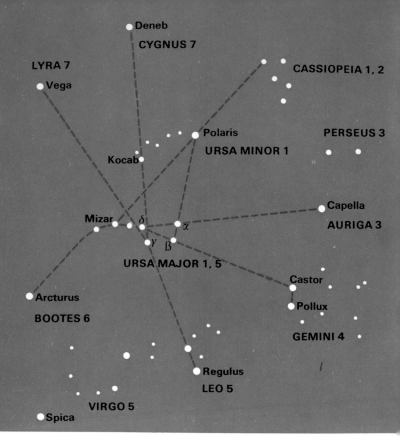

Key map 1 – Ursa Major, the Great Bear, and its relationship to other constellations

cuous and unmistakable. The Great Bear ('Ursa Major' in the Latin form) can be seen from anywhere which lies north of latitude 40 degrees south, whilst Orion straddles the celestial equator and can thus be seen from anywhere in the world. Guide lines for finding the other important constellations from Ursa Major and Orion are shown on these charts.

Since there is no way of representing a sphere completely accurately on flat paper, there is no perfect way of mapping the stars. There is always a small degree of distortion, but this does not greatly affect the maps which follow. The sky has been divided into north and south polar areas – that is, regions lying north of declination 50 degrees north and south of 50 degrees south – and the middle heavens between these two zones. The middle heavens have been divided into six parts, each rather more than 4 hours of right ascension wide (see page 77) to give good overlap with its neighbours. Each one of these separate maps, if it does not contain one of the key constellations, contains at least one of the constellations easily found from one of the key constellations, and thus, with patience, the whole sky can be filled in.

On each map there is a grid of lines representing right ascension and declination to give a guide to the celestial coordinates of the constellations. Beside each chart is the time of year at which the centre part of the map is due south at midnight (local time). Remember that the constellations rise two hours earlier each month. Stars which lie farther south in declination than 90 degrees minus your own latitude (in the northern hemisphere) can never be seen. If you live in the southern hemisphere, the situation is reversed.

Key map 1 – The Great Bear

The seven brightest stars of this famous constellation form that part often known as the Plough or Big Dipper. Four stars form the ploughshare and the other three the handle. This group is circumpolar north of 40 degrees north.

A line through the stars ß and α, known as the 'pointers', leads to Polaris, the pole star, which is a 2nd magnitude star lying some 30 degrees distant, pretty well on its own. (A useful thing to remember is that the spread of a hand at arm's length corresponds roughly to about 20 degrees of arc.) Polaris lies at the tail end of the faint constellation Ursa Minor. If we extend this line we come to Cepheus. A line drawn from Mizar through Polaris for a further 30 degrees leads to the conspicuous W-shaped group of stars forming Cassiopeia. Extending the curve of the tail of the Plough downwards for some 30 degrees leads to Arcturus, the brightest star in Boötes and a further 30 degrees distant is Spica in Virgo. Leo lies in line with δ and γ Ursae Majoris. Other lines shown lead to Lyra, Cygnus, Auriga and Gemini.

Along with the constellation names on the key maps are given the numbers of the other maps on which they appear.

Key map 2 – Orion

Orion is possibly the most striking constellation in the heavens. His outline is easy to see, with his 'sword' hanging from the three stars which make up his 'belt'. Orion boasts two 1st-magnitude stars, the red Betelgeuse and the blue-white Rigel.

A line through the belt leads up to Aldebaran in Taurus, about 20 degrees distant, and thence to the Pleiades cluster. About the same distance from the belt in the other direction lies Sirius, the brightest star in the sky, in the constellation Canis Major. Procyon, in Canis Minor, is readily found. A line through κ, the belt, and λ leads to Auriga and its bright star Capella, some 40 degrees away, whilst a line through Rigel and Betelgeuse leads to the two stars Castor and Pollux in Gemini, the Twins. Leo lies some 60 degrees distant on a line from κ to Procyon. Cetus, the Whale, is an ill-defined constellation found by taking a line from Betelgeuse through γ, while Hydra lies beyond Procyon on a line from Betelgeuse.

Key map 2 – Orion and its relationship to other constellations

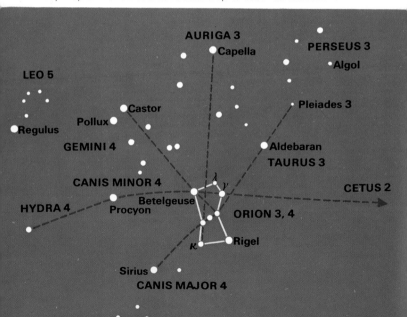

Map 1 – North polar stars

The north polar constellations are easy to learn.

Ursa Major is a key constellation readily recognizable by the seven stars which form the Plough. However, the group spreads over a much larger area than this, extending in broad sweeps ahead of and below the Plough as shown on maps 5 and 6. Most of the stars are of 2nd magnitude, and Mizar, the middle star of the handle, is a double, its companion Alcor being readily visible to the naked eye. In fact, telescopically, the system is seen to be triple. The constellation contains M81, a particularly beautiful galaxy.

Ursa Minor. The seven principal stars of this small and rather faint constellation form a shape similar to the Plough except that the tail curves in the opposite direction. Polaris, the pole star, lies at the end of the tail and is, in fact, about one degree from the true celestial pole. Of magnitude 2·1, Polaris is a double star with a 9th magnitude companion visible in a three-inch refractor. Curiously enough, the four stars forming the square are of magnitudes 2, 3, 4 and 5.

Draco. A long, faint constellation twisting around Ursa Minor, Draco depicts a dragon. One of the stars in the tail of the Dragon was the pole star at the time of the building of the pyramids. The constellation contains a planetary nebula visible in amateur-sized telescopes.

Cepheus. This constellation is not too easy to identify and can best be recognized by thinking of it as a little house with a peaked roof. δ Cephei, which varies in magnitude from 3·7 to 4·3 in just under five and a half days, is the prototype of the Cepheid variables (see page 85) and lies near the bottom left corner of the 'house'. Near the centre of the base of the 'house' lies an irregular variable of an astonishing garnet colour.

Cassiopeia. A very notable constellation shaped like a W, it depicts a queen seated on her throne. The Milky Way runs through the constellation and the group presents a beautiful sight through binoculars. There are many double stars and two fine clusters. In 1572 a supernova explosion occurred within the constellation, the star concerned outshining even Venus for a time before fading from view.

Camelopardalis. A large and very dull constellation.

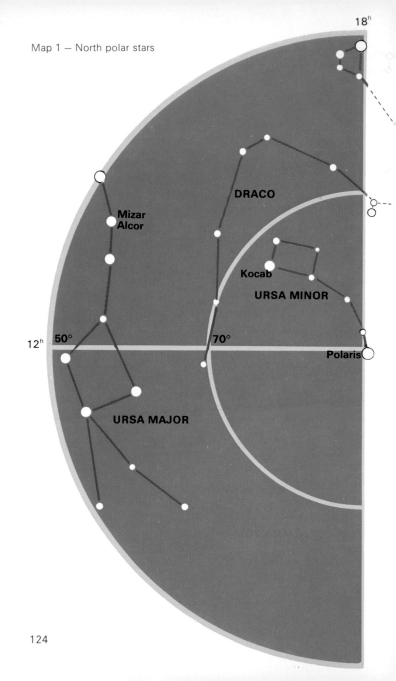

Map 1 — North polar stars

18^h

DRACO

Mizar
Alcor

Kocab

URSA MINOR

12^h 50°   70°

Polaris

URSA MAJOR

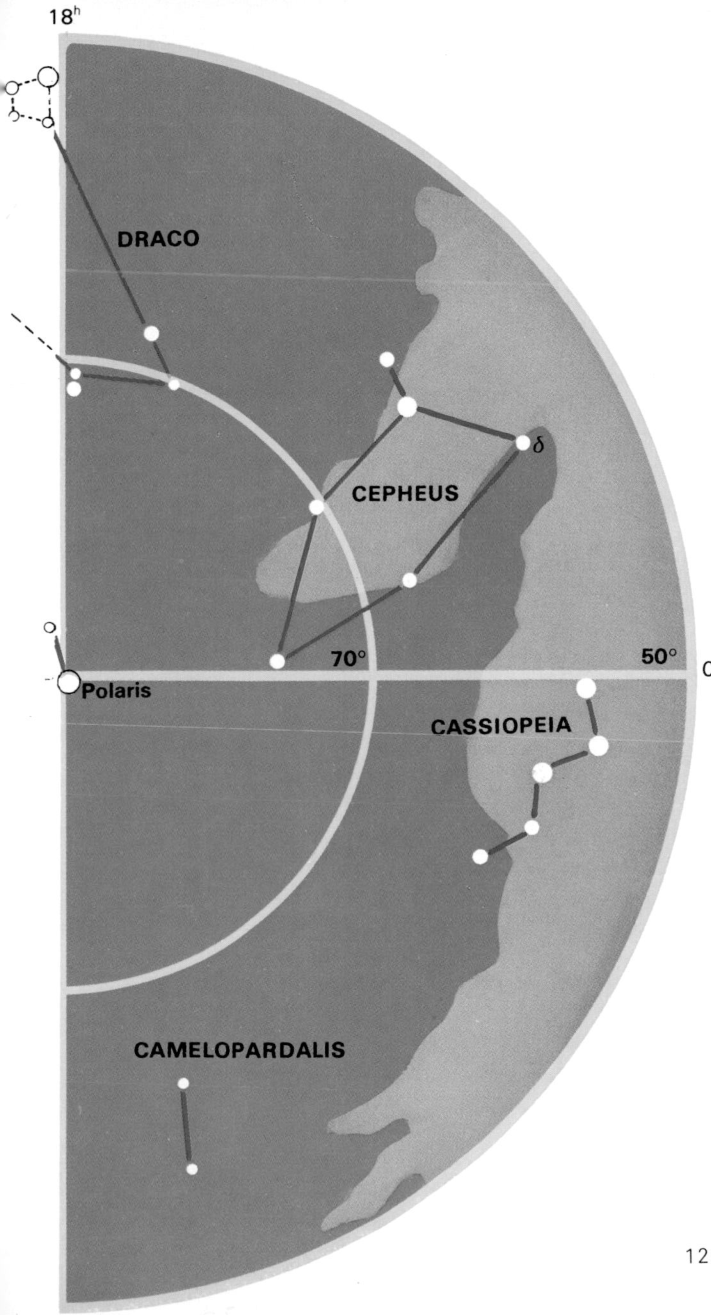

18^h

DRACO

CEPHEUS

δ

70°

50°

0^h

Polaris

CASSIOPEIA

CAMELOPARDALIS

125

Map 2 – Right ascension 22 hours to 2 hours

Pegasus. A line from the pole star through Cassiopeia extended for some 30 degrees leads to Pegasus, the Winged Horse. The constellation is easily identified by the four principal stars which form a square of about 15 degrees on each side, though none of the stars is brighter than 2nd magnitude. The star on the top left corner of the square as seen in the northern hemisphere is actually α Andromedae and forms the point of contact between Pegasus and Andromeda. ß Pegasi, on the top right corner, is variable by about half a magnitude.

Andromeda, daughter of Cassiopeia, extends from the corner of the square of Pegasus, the principal three stars all being of 2nd magnitude. The third of these, counting from α, is γ, a fine double star with yellow and blue components. Extending from ß at right angles to the line of α, ß and γ are two faint stars, and beyond these lies the famed M31, the Great Nebula in Andromeda. Just visible to the naked eye, this is in fact a galaxy very similar to our own, though rather larger, and at just over 2 million light years away, it is one of the nearest external galaxies.

Pisces. Representing the Fishes, Pisces is one of the constellations of the Zodiac (see page 142), though a rather faint one. Owing to precession, it now contains the First Point in Aries, which is the point at which the Sun crosses the celestial equator going northwards in the northern hemisphere Spring. Needless to say, this point used to be in the constellation Aries.

Cetus is a large and ill-defined constellation representing the Whale. The only bright star is ß, of 2nd magnitude, though there are several 3rd-magnitude stars.

Aquarius. Also a Zodiacal constellation, Aquarius has no stars brighter than 3rd magnitude. It does, however, contain a fine group of orange stars, nicely seen in binoculars.

Piscis Australis. This small group contains Fomalhaut, a fine 1st-magnitude star, but all the other stars are fainter than 4th magnitude.

Phoenix contains one 2nd-magnitude and two 3rd-magnitude stars, one of which is a double.

Grus. Easily seen, Grus contains two 2nd-magnitude stars.

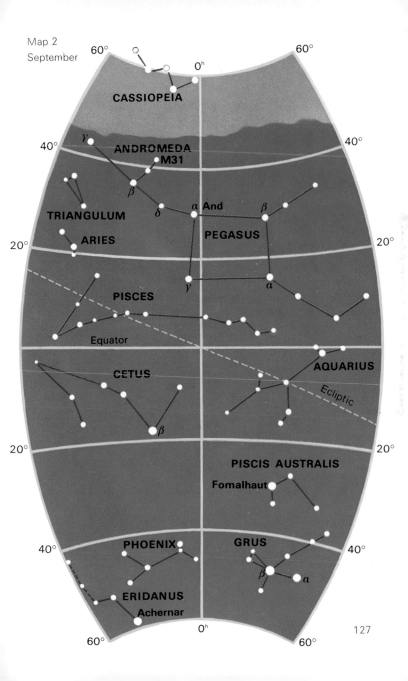

Map 2
September

60°   0ʰ   60°

CASSIOPEIA

40°   ANDROMEDA   40°

γ

M31

β

δ   α And   β

TRIANGULUM   PEGASUS

20°   ARIES   20°

γ   α

PISCES

Equator

CETUS   AQUARIUS

20°   Ecliptic   20°

β

PISCIS AUSTRALIS

Fomalhaut

40°   PHOENIX   GRUS   40°

β

α

ERIDANUS

Achernar

60°   0ʰ   60°

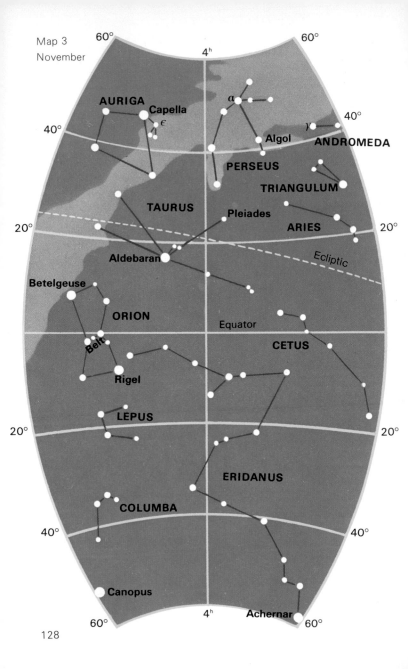

Map 3
November

60°
4ʰ
60°

AURIGA

Capella

ε

40°
40°

α

Algol
ANDROMEDA

PERSEUS

γ

TRIANGULUM

TAURUS

Pleiades

ARIES

20°
20°

Aldebaran

Ecliptic

Betelgeuse

ORION

Equator

CETUS

Belt

Rigel

LEPUS

20°
20°

ERIDANUS

COLUMBA

40°
40°

Canopus

60°
4ʰ

Achernar
60°

128

Map 3 – Right ascension 2 hours to 6 hours

Auriga. This, the Charioteer, is a very conspicuous constellation and contains the yellow 1st-magnitude star Capella (actually magnitude 0·1). Nearby is the giant eclipsing binary ε. Lying on the edge of the Milky Way, the group contains some fine clusters.

Perseus. Lying near Cassiopeia, this is an interesting constellation lying on the Milky Way. It contains Algol, the prototype eclipsing binary which varies from magnitude 2·2 to 3·4 and is described on page 84. At the 'Cassiopeia end' of the group lies the magnificent double cluster, visible to the naked eye and breathtaking when seen through binoculars.

Taurus. Depicting the Bull, this constellation contains the red, 1st-magnitude Aldebaran, a double star with a faint companion. Around Aldebaran lies the fine V-shaped cluster of the Hyades and, beyond, the famed Pleiades cluster. Most observers can see seven stars in this group with the naked eye, but some can see a dozen or more. There are several hundred stars in all in the cluster. Also in Taurus lies the Crab Nebula, the remnants of a supernova.

Aries. An inconspicuous constellation representing the Ram, with, however, one 2nd-magnitude star.

Triangulum. A small triangular constellation lying near Andromeda, this group contains a faint spiral galaxy M33.

Orion. This spectacular constellation is one of the key constellations and boast of two 1st-, five 2nd-, and four 3rd-magnitude stars. The red Betelgeuse is a variable supergiant with average magnitude 0·6, whilst Rigel is a blue-white giant of magnitude 0·2 and is, in fact, a double, being easily separated in a 3-inch refractor. Below the three stars of the belt hangs the faint sword which contains a host of interesting objects including the Great Nebula M42 which is a mass of glowing gas some 26 light years across and about 1,600 light years distant. It can be seen with the naked eye as a misty patch and contains a quadruple star, the Trapezium.

Lepus, the Hare, is a small group south of Orion.

Columba represents the dove which flew from Noah's Ark.

Eridanus. A long twisting constellation containing the 1st-magnitude star Achernar.

Map 4 – Right ascension 6 hours to 10 hours

Lynx. This is a large constellation named by Hevelius. It contains no conspicuous stars.

Gemini. Representing the Twins, this is a fine, conspicuous constellation readily distinguished by the two close and similar stars Castor and Pollux (magnitudes 1·6 and 1·1 respectively). They are in fact 5 degrees apart and form a useful measuring guide. Castor is an easily resolved double in small telescopes, but in fact consists of three components, each of which is itself a close double. Near the right foot of the Twins lies a very fine open star cluster.

Cancer. A faint constellation, the Crab is vaguely similar in outline to Orion. It is chiefly noted for the very fine open cluster M44, known as Praesepe, the Beehive, which is visible to the naked eye as a misty patch.

Canis Major. Representing one of Orion's two dogs, this constellation contains the brightest star in the sky, Sirius, of magnitude −1·4 and lying only 8·6 light years away. Sirius is of spectral class A1 (see page 88) and is 26 times as luminous as the Sun. The cluster M41 is a fine sight in binoculars.

Canis Minor. The lesser of the two dogs, this group also contains a 1st-magnitude star, Procyon (magnitude 0·4) which is about 10 light years distant. It contains nothing else of interest.

Hydra. The largest constellation in the sky, Hydra winds its way across the heavens. However, it is a faint and rather uninteresting group. α, the brightest star, is of 2nd magnitude but is easily found on account of its solitary position; it was named Alphard, 'the solitary one', by the Arabian astronomers.

Monoceros. A faint constellation, this contains a 5th-magnitude triple star and a pleasing open cluster.

Puppis and Vela are two components of the constellation Argo Navis, the Ship, which has now been broken up into three parts, the other being Carina. Puppis represents the poop of the ship and contains a fine cluster. Vela is the ship's sail and has four 2nd-magnitude stars. Carina is described on map 8.

Leo is described on map 5.

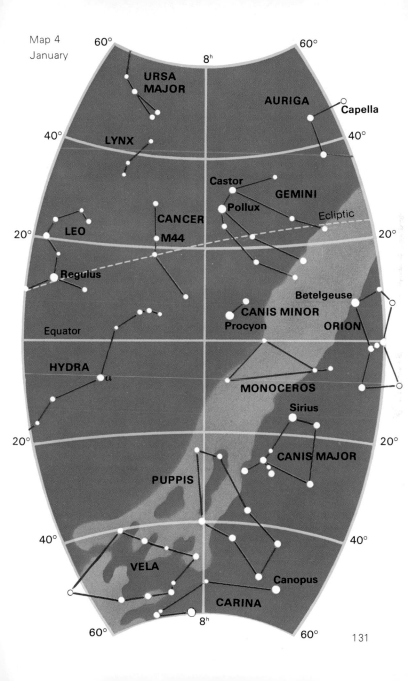

Map 4
January

60° 8ʰ 60°

URSA MAJOR

AURIGA ○ **Capella**

40° 40°

LYNX

Castor

GEMINI

CANCER Pollux

Ecliptic

20° **LEO** 20°

M44

Regulus

Betelgeuse ○

CANIS MINOR **ORION**

Procyon

Equator

HYDRA α

MONOCEROS

Sirius

20° **CANIS MAJOR** 20°

PUPPIS

40° 40°

VELA

○ Canopus

CARINA

60° 8ʰ 60°

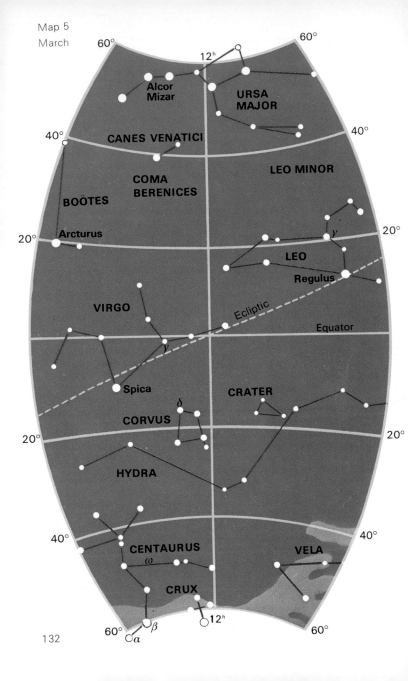

Map 5 – Right ascension 10 hours to 14 hours

Canes Venatici and Coma Berenices are adjacent faint constellations but the whole area is littered with clusters and nebulae and is well worth sweeping with binoculars. The principal star of Canes Venatici is a wide optical double of magnitude 2·9 and it contains M3, a very fine globular cluster. Coma Berenices contains no stars brighter than 4th magnitude.

Leo, the Lion, is the most conspicuous of the Zodiacal constellations, the most notable feature being the curved sickle of stars beginning with the blue-white 1st-magnitude Regulus. γ Leonis, the third star of the Sickle, is a fine double, magnitudes 2·4 and 3·8. The constellation forms the radiant for the famed Leonid meteor shower, which appears around 17 November and in 1833, 1866 and 1966 produced spectacular displays with thousands of meteors per hour.

Leo Minor. Lying to the north of Leo, this small, faint group contains nothing of great interest.

Virgo, the Virgin, is shaped rather like a Y with the brightest star, Spica, at the base of the Y. The constellation continues further in increasing right ascension but this part is rather faint. Spica is a blue-white star of exactly magnitude 1 and is 1,500 times brighter than the Sun. There are many faint galaxies in this region, some of which are visible in small telescopes. γ is one of the finest binary stars for small telescopes and is the first 3rd-magnitude star in the direction of Leo from Spica.

Corvus. This small constellation representing the Crow is easy to find as its four principal stars are all about 3rd magnitude and form a compact quadrilateral. δ is double.

Hydra. This constellation continues to wind its way across this part of the sky. This part contains a red irregular variable.

Centaurus. The Centaur is a bright, striking group, containing two 1st-magnitude stars α and ß. α is a superb double star and also contains within its system a faint red dwarf star, Proxima Centauri, the nearest star to the solar system at a distance of $4\frac{1}{4}$ light years. It lies very close to α. The constellation also contains, near its centre, ω Centauri, the finest of the globular clusters.

Map 6 – Right ascension 14 hours to 18 hours

Hercules. This is a large constellation but contains no stars brighter than 3rd magnitude. Four stars in a rough square form the centre of Hercules, from which the arms and legs radiate. The constellation contains the magnificent globular cluster M13 which can just be seen with the naked eye on a clear moonless night. The cluster is some 34,000 light years away and contains 50,000 stars. A less prominent globular cluster M92 lies 10 degrees away.

Corona Borealis, the Northern Crown, is easily recognizable as such and is a compact but interesting constellation with several double and variable stars.

Boötes, the Herdsman, is a fairly well defined constellation containing the orange 1st-magnitude star Arcturus (actual magnitude 0·1) and lies near the Great Bear which he is supposed to be pursuing. δ Boötis is a double which is very easily resolved in a small telescope and lies adjacent to Corona Borealis.

Ophiuchus. Although it contains several stars of 2nd and 3rd magnitude, this constellation is not too easy to identify and contains little of interest with the exception of M19, a fine globular cluster.

Serpens Caput, the Head of the Serpent, also contains a fairly bright globular cluster, M5, which lies near α, the brightest star in the constellation (magnitude 2·7).

Libra. Depicting weighing scales, Libra is a rather dull Zodiacal constellation. δ Librae is an eclipsing binary with period of 2·3 days and a magnitude range of 4·8 to 6·2. Its distance is 205 light years.

Scorpius represents the scorpion which bit Orion. A splendid constellation, the head and long curving tail are easy to distinguish. The principal star is Antares, a deep red supergiant of magnitude 0·98 with a diameter nearly 300 times greater than the Sun. Antares has a 7th-magnitude green companion. The fine globular cluster M80 lies near Antares, as does the faint open cluster M4.

Ara. This constellation contains three 3rd-magnitude stars but nothing of interest.

Lupus contains a large number of 3rd-magnitude and faint stars but is not notable.

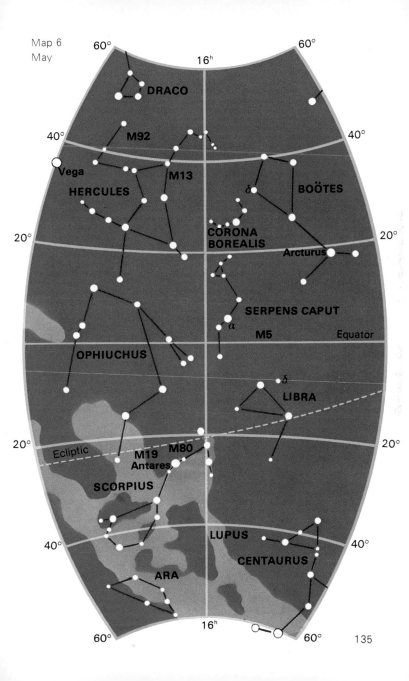

Map 6
May

60° 60°

16ʰ

DRACO

40° M92 40°

Vega

M13

HERCULES BOÖTES

δ

CORONA
BOREALIS

20° 20°

Arcturus

SERPENS CAPUT

α

M5 Equator

OPHIUCHUS

δ

LIBRA

20° 20°

Ecliptic

M19 M80

Antares

SCORPIUS

LUPUS

40° CENTAURUS 40°

ARA

60° 60°

16ʰ

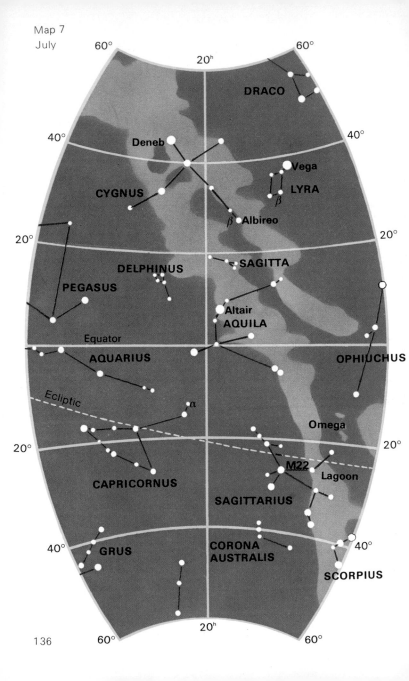

Map 7
July

60° 60°

20ʰ

DRACO

40° 40°

Deneb Vega

LYRA

CYGNUS β

Albireo
20° β 20°

SAGITTA

DELPHINUS

PEGASUS Altair
AQUILA

Equator

AQUARIUS OPHIUCHUS

Ecliptic

α

Omega

20° 20°

M22
CAPRICORNUS Lagoon

SAGITTARIUS

40° 40°

GRUS CORONA
AUSTRALIS
SCORPIUS

20ʰ

60° 60°

Map 7 – Right ascension 18 hours to 22 hours

Cygnus, the Swan, is a beautiful constellation, often known as the Northern Cross because of its shape. The brightest star, Deneb, is a white 1st-magnitude star lying in the head of the swan. At the swan's tail lies ß Cygni, Albireo, a 3rd-magnitude yellow and green double, one of the finest doubles in the entire sky. 61 Cygni, a 5th-magnitude star, was the first to have its distance measured (10·6 light years). There are many clusters in this area, the Milky Way running through the constellation.

Lyra, the Lyre, is a small constellation but full of interesting objects. Vega, the only star in the group brighter than 3rd magnitude, is blue-white and some 26 light years distant. Besides Arcturus, it is the brightest star in the northern hemisphere (magnitude 0·04). ε Lyrae, a 4th-magnitude star close to Vega, is the famous double double easily separated in a 3-inch refractor (see page 83). ß Lyrae, at the bottom right corner of the diamond-shaped part of the group, is an eclipsing binary. Nearby is the Ring Nebula, a well known planetary nebula.

Sagitta is a compact arrow-shaped group.

Delphinus. Lying on a line from Vega through ß Cygni this neat little group contains a fine 5th-magnitude double star. An amateur observer, George Alcock, discovered a nova in this constellation in 1967, a nova which persisted for an exceptionally long time.

Aquila. Following the Swan up the Milky Way, Aquila, the Eagle, is a fine constellation. The chief star, Altair, is a white 1st-magnitude star, and together with Deneb and Vega forms a very fine triangle, often known as the 'Summer Triangle'. The group contains a 4th-magnitude Cepheid.

Capricornus is a rather dull Zodiacal constellation with no stars brighter than 3rd-magnitude. α Capricorni, at the 'Aquila end' of the group, appears from Earth as a naked-eye double. In fact, the two stars are not in close proximity. One of the stars is a giant, the other a super-giant.

Sagittarius is a large, bright group on a line through Deneb and Altair. It contains the Omega Nebula, the Lagoon Nebula, and the globular cluster M22, all well seen in medium telescopes.

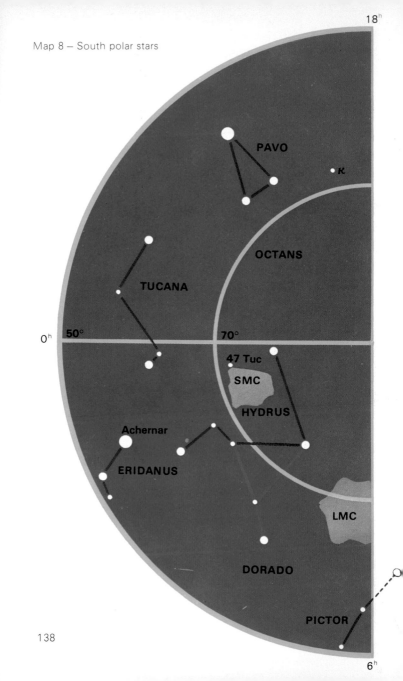

Map 8 – South polar stars

18ʰ

PAVO

κ

OCTANS

TUCANA

0ʰ 50° 70°

47 Tuc

SMC

HYDRUS

Achernar

ERIDANUS

LMC

DORADO

PICTOR

6ʰ

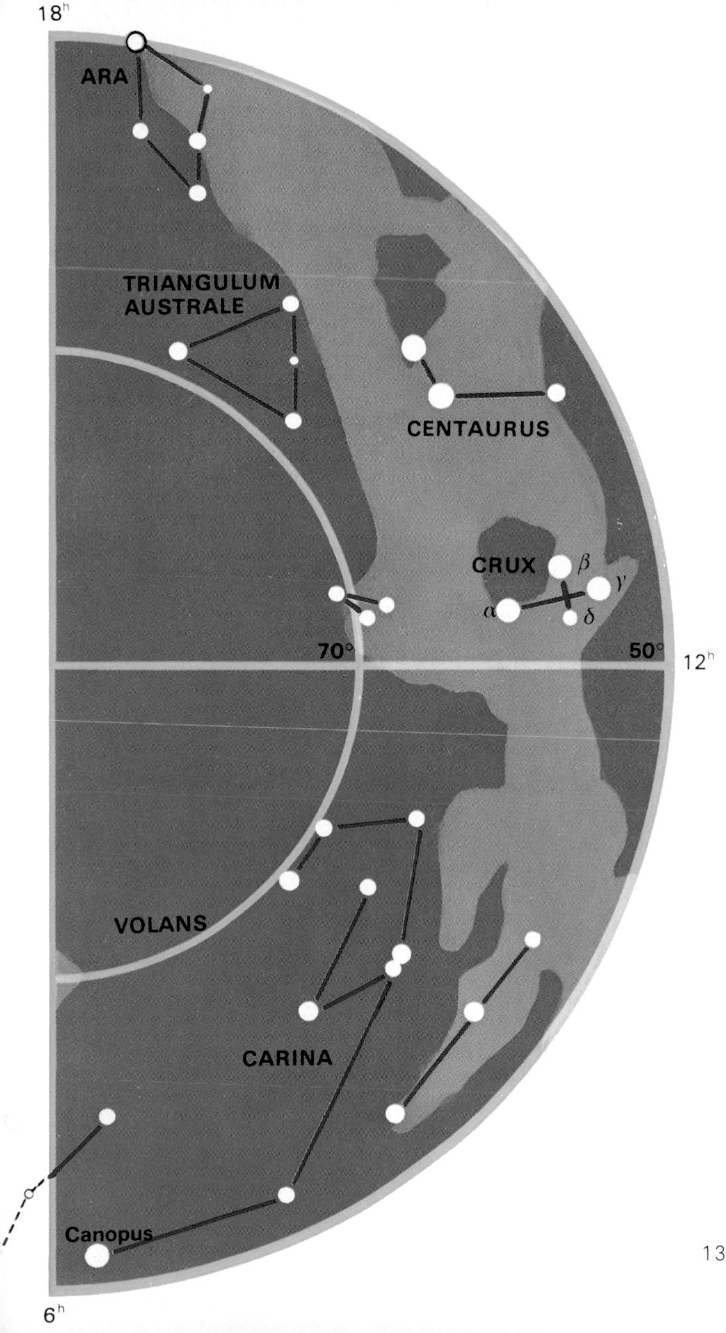
18h
ARA
TRIANGULUM
AUSTRALE
CENTAURUS
CRUX
β
γ
α
δ
70°
50°
12h
VOLANS
CARINA
Canopus
6h

Map 8 – South polar stars

The observer living in the southern hemisphere has to remember that the constellations in the preceding maps are drawn 'upside down' from his point of view. However, the maps can be used upside down to get round this. The south polar groups on the following map present no problem. Unfortunately, the south polar area is not as tidy as the north.

Crux, the Southern Cross, is the best known of the southern constellations. α and ß are both 1st-magnitude, while γ is of magnitude 1·6. γ and α point close to the south celestial pole, some 30 degrees distant. α is a fine double. Near β is a magnificent open cluster of over 100 stars.

Centaurus is described on map 5.

Triangulum Australe is a bright little group containing a fine open cluster.

Pavo. A faint group apart from one 2nd-magnitude star. κ is a Cepheid variable, fluctuating between magnitudes 4·0 and 5·5 in nine days.

Tucana, the Toucan, is a faint constellation but does contain most of the Small Magellanic Cloud (see page 141) as well as the very beautiful open cluster 47 Tucanae and a globular cluster. Both these clusters can just be seen with the naked eye. The whole area repays study with binoculars.

Octans. This faint constellation contains the south celestial pole. Unfortunately, there is no bright star near the pole. In fact the nearest 2nd-magnitude star is over 20 degrees away.

Hydrus. A twisting constellation of little interest.

Dorado. The brightest star in this constellation is of magnitude 3·5, but it does contain a large part of the Large Magellanic Cloud, as well as the Great Looped Nebula.

Volans. This small group contains two nice 4th-magnitude doubles.

Carina, the hull of the ship Argo, is a large, striking constellation. Canopus, a yellow star of magnitude $-0·7$, is the second brightest star in the sky and lies 650 light years away, shining 80,000 times brighter than the Sun. The group contains a large globular cluster.

The Magellanic clouds

The Magellanic Clouds, named after the navigator and explorer Ferdinand Magellan, who first recorded them on his circumnavigation of the world in 1519, are two small galaxies, the closest to our own. They are clearly visible to the naked eye in the southern hemisphere, looking rather like detached portions of the Milky Way. The Large Cloud, or Nubecula Major, is some 8 degrees in diameter and is visible even in the presence of moonlight, whilst the Small Cloud, Nubecula Minor, is fainter and about half the size. Unfortunately neither can be seen at all anywhere north of latitude 15 degrees north.

They are both irregular galaxies, though the larger does show indications of a barred spiral form (see page 101), and they lie only 160,000 light years distant. Both are much smaller than our galaxy, being about 25,000 (Large Cloud) and 10,000 light years in extent, and are classed as satellite galaxies of our own. The Andromeda galaxy has similar satellites (page 100). Both Clouds contain many Cepheid variables, and the Large Cloud has a large proportion of blue giants.

The Large Magellanic Cloud

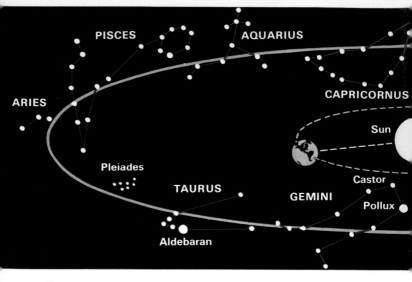

The constellations of the Zodiac

The Zodiac

As the Earth moves on its path around the Sun, the Sun is projected against the background constellations and follows a path known as the *ecliptic*. The belt of sky, about 18 degrees wide, centred on the ecliptic, is known as the Zodiac, and the constellations in this belt are the Zodiacal constellations. The Moon and planets all move within this band of sky.

The twelve constellations of the Zodiac are as follows: Aries, the Ram (described on map 3); Taurus, the Bull (map 3); Gemini, the Twins (map 4); Cancer, the Crab (map 4); Leo, the Lion (map 5); Virgo, the Virgin (map 5); Libra, the Scales (map 6); Scorpius, the Scorpion (map 6); Sagittarius, the Archer (map 7); Capricornus, the Goat (map 7); Aquarius, the Water Carrier (map 2); and Pisces, the Fishes (map 2). Starting at the first point in Aries, or *vernal equinox*, where the ecliptic intersects the celestial equator in the northern hemisphere Spring, the Zodiac is divided into twelve zones 30 degrees wide. Each of these zones is assigned a 'sign of the Zodiac' corresponding to one of the twelve con-

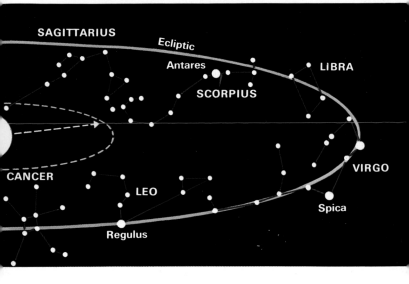

stellations named above, but the period during which the Sun lies in a particular sign does not correspond well with the time that it lies in the actual constellation of that name. The reasons for this are that the constellations are not of the same width, nor are they evenly spaced, and – most important – precession (page 9) has caused the vernal equinox to move from Aries into Pisces. In a thousand years' time it will lie in Aquarius, and so it will move through the Zodiac, returning more or less to its present position in about 26,000 years.

The motion of the Sun through the Zodiac enabled the ancients to estimate the seasons and regulate their crops, and the Zodiac came to assume a position of prime importance in the study of astrology. Naturally enough in ancient times, the motions of the Sun, Moon and planets against the background of the Zodiac were thought to have profound influence on the course of man's affairs, and the influence on individuals was governed by the sign under which they were born. Astrology still claims some adherents today, but it has no scientific basis. Astronomers become most annoyed if confused with astrologers!

THE AMATEUR ASTRONOMER

Unlike most other sciences, the amateur can still play an important part in astronomy. There are many fields, such as variable stars, seeking comets and novae, observing changes on planetary surfaces (particularly Jupiter), in which serious amateurs are contributing very useful work, particularly as these are fields which most professional astronomers do not have time to study. Most observatories are mainly concerned with long range photographic and spectroscopic observations of stars and galaxies – and quite rightly too – and little time is available for such long drawn-out series of observations as are required for irregular variables, for example.

However, one of the greatest joys of astronomy is simply 'just looking' and the sky has a great deal to offer the observer without any sort of telescope at all. At first the sky seems a confused mass of stars, but with very little practice it be-

Conjunction of the Moon, Venus and Jupiter

comes possible to pick out the major constellations and then to fill in the less obvious ones. And there is always the chance – remote though it may be – of being the first to detect a nova which has suddenly flared up to naked-eye visibility. If you know the constellations well, a new star stands out very clearly. Recognizing the planets and following their movements can be a source of considerable pleasure, particularly detecting the elusive Mercury or the very faint Uranus. Meteors are best seen with the naked eye, and, armed with a good watch and a star map, an amateur can usefully plot their behaviour. The night sky is always changing, with eclipses, occasional comets and many other features, and there will always be something to arouse new interests and excitement.

A pair of good binoculars will open up fascinating new possibilities. Many more stars can be seen, doubles and multiples, clusters and nebulae, as well as craters on the Moon, the moons of Jupiter, the phases of Venus, and so on. Set up a deck chair, wrap up well (it gets cold after a time), settle back with a pair of binoculars and sweep along the Milky Way. It is quite a sight!

A simple altazimuth mounting can be easily made from wood. The overhang allows the telescope to be pointed vertically upwards. A very useful binocular size for astronomy is 7 x 50 – i.e. magnification of 7 times with 50 millimetre object glasses.

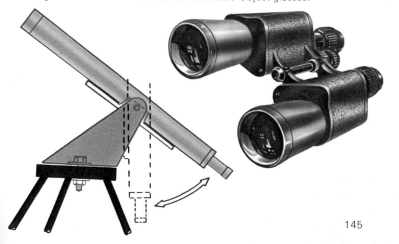

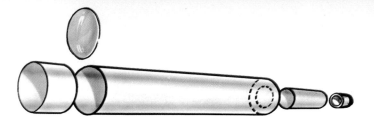

A simple refractor using cardboard tubing

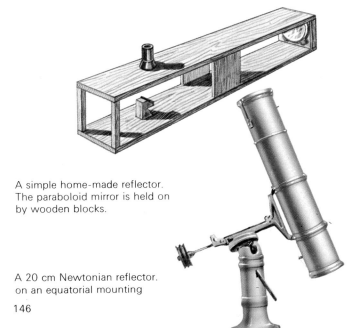

A simple home-made reflector.
The paraboloid mirror is held on
by wooden blocks.

A 20 cm Newtonian reflector.
on an equatorial mounting

Amateur telescopes

It is a very simple matter to construct a rudimentary refractor of, say, 30 to 50 times magnification, which will show sunspots well and the Moon in detail. An optician should be able to supply an ordinary convex lens, say 5 cm in diameter and of 1 metre focal length, and also a smaller lens with a focal length of 2 cm, at very little cost. All that is needed now is a piece of cardboard tubing about 1 metre long and 5 cm inside diameter, and a short piece, say 15 cm long, which will slide into the larger one. Mount the 5 cm lens, the object glass, at one end of the long tube, and fix the other lens, the eyepiece, in the short tube, which can then be adjusted by sliding it in and out of the larger tube until a well-focused image is seen. Since the magnification of a telescope is given by

$$\frac{\text{focal length of object glass}}{\text{focal length of eyepiece}}, \text{ this telescope will magnify } \frac{100 \text{ cm}}{2 \text{ cm}} = 50 \text{ times.}$$

Of course, such a telescope will suffer from chromatic aberration, but it will be very useful none the less. To get a more powerful instrument, it is better to construct a reflector, and many amateurs have built reflectors up to 30 cm aperture and larger. However, a 15 cm or 20 cm reflector is the most common size, and the Newtonian pattern is nearly always employed for simplicity. Mirrors of suitable size (usually focal ratios of about f/8) are made by several optical companies, but with care and patience it is a relatively simple task to make your own main mirror (see bibliography). The small flat mirror and eyepieces are best bought; but even so, if you make your own mirror, a complete 15 cm reflecting telescope can be built and mounted at very little expense.

To be of any value, a telescope must be well mounted, preferably on some sort of equatorial such as the German (page 24), the yoke or the fork mount. 7·5 cm refractors and 15 cm reflectors can perform useful work, but a 10 cm or 15 cm refractor or 20 cm reflector is really very useful indeed and many well known amateurs have telescopes of this size.

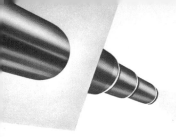

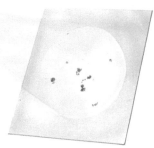

The Sun's image can be projected on a white screen with a card screen around the telescope barrel to make a shadow. Sunspots can be seen in this way.

There are many useful additions that the amateur can make to his telescope. The *Barlow lens*, for example, is a small lens placed in front of the eyepiece which increases the focal length of the telescope by variable amounts, and so different magnifications can be attained with the same eyepiece.

To simplify observing, it is useful to have the polar axis of the telescope mount driven at the exact speed necessary to counteract the Earth's motion. This is quite feasible with, for example, a synchronous electric motor geared down to give one rotation in 23 hours 56 minutes. At any rate, some sort of hand operated 'slow motion' mechanism is necessary if high magnifications are used.

Astronomical photography is not too difficult either, since the only 'camera' needed, basically, is a box to hold the film, with a simple shutter. This is simply clamped to the eyepiece. The only difficult operation is focusing the image onto the film. If, of course, you happen to have a single-lens reflex camera, then you can do a great deal.

Amateur observations

Let us see what the amateur can do in various fields.

The Sun. A daily record of the development and numbers of sunspot groups is useful, and with plenty of light available, photography has many possibilities. The amateur W. Baxter regularly produces sunspot photographs, using a 10 cm refractor, which are as good as the best available. The most

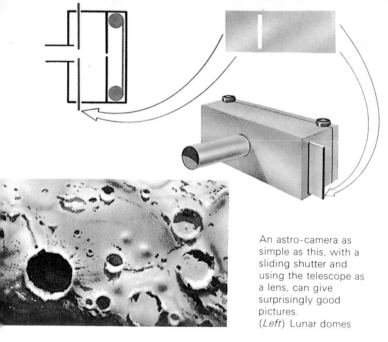

An astro-camera as simple as this, with a sliding shutter and using the telescope as a lens, can give surprisingly good pictures.
(*Left*) Lunar domes

convenient, and the safest, way to observe the Sun is by projection, as shown.

The Moon. At one time lunar observation was almost entirely the province of the amateur, but space exploration has changed this. The mapping of lunar features is of little value now, except for interest. However, special features, such as domes and dark bands in craters can be charted, and observations of transient phenomena, coloured glows and the like, are very valuable. A 20 cm telescope will show tremendous detail.

Venus. Very little can be made out on the cloud layer which covers the planet, except a few hazy patches which should be recorded if possible; likewise the bright patches which appear sometimes near the tips of the crescent, known as *cusp caps*. An interesting thing about Venus is the phase anomaly or *Schröter effect*. The stage when Venus is at half-moon phase is known as *dichotomy* and should occur exactly when predicted. However, dichotomy never seems to occur quite on time and observations of this are useful.

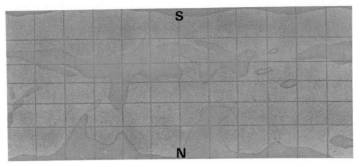

Map of Mars on Mercator's projection

Mars. Many surface features can be seen, and detailed drawings of their development, colours, etc., are required, as well as observations of the polar caps, clouds and such features. A set of colour filters is useful (and these can be used too on lunar work). However, Mars presents a small disc, and fairly large telescopes are required for really useful work (25 cm or so). Nevertheless, Mars is a fascinating world.

Jupiter. The giant planet is the most fruitful for the amateur – plenty of detail is visible with a 15 cm reflector and the cloud belts are always in turmoil, sometimes changing quite dramatically. What are required are *transit observations,* that is, observations of the times, to the nearest minute, that various spots, humps, hollows, etc., in the cloud belts pass the central point (the central meridian) of the planet.

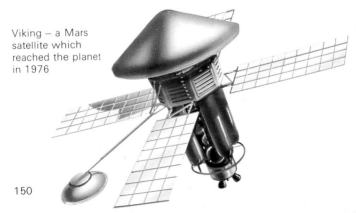

Viking – a Mars satellite which reached the planet in 1976

Reference to tables, such as the *BAA Handbook* (see page 155), will give the longitude of these features, and so their motions can be traced.

Saturn. The ringed planet is the most beautiful, but few details can be seen on the surface. However, when they do appear, they should be recorded in the same fashion as with Jupiter. The great white spot which appeared on the planet in 1933 was discovered by an amateur, W. T. Hay. Observations of the intensities of the rings and their shadows too are needed.

All planetary observations should have recorded the date, time, name of observer, size and type of telescope, magnification, and some indication of what observing conditions were like.

Variable stars. This is a very good field for the amateur. Observations of the brightness variations of long-period and irregular variables in particular are left largely to amateurs. The technique for brightness estimation involves comparing the variable with several comparison stars of known magnitude similar to that of the variable. Several systems have

The belts of Jupiter   The rings of Saturn

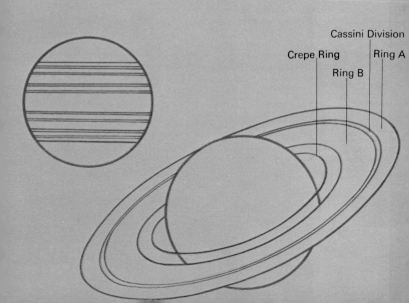

Cassini Division
Crepe Ring   Ring A
Ring B

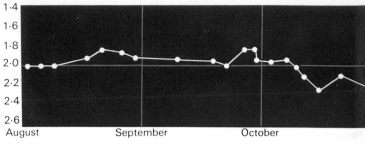

1·4			
1·6			
1·8			
2·0			
2·2			
2·4			
2·6			
August	September	October	

The light curve of an irregular variable

been evolved: basically they consist of estimating the variable's brightness either by mentally estimating steps of a tenth of a magnitude from the comparison star to the variable, or by estimating the fractional difference in brightness between the variable and comparison stars. For example, a variable might lie, in intensity, one third of the way from one star to another. It is important that these observations be carried out over long periods, when a graph of the intensity variations can be plotted. Some variables can be followed by the naked eye, many more with binoculars or telescopes.

Novae. Systematic searches of the sky with binoculars or telescopes (or even the naked eye) can, with a great deal of luck, lead to the detection of novae. Photographic studies are rather expensive. The chances of finding one are very slim, but a discovery is of the utmost importance. Position and brightness should be carefully noted.

Comets. These can be searched for in a similar way as novae. Careful sweeping of the sky is essential, and the

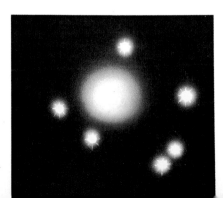

Alcock's First Comet of 1959, discovered by the English amateur G. E. D Alcock

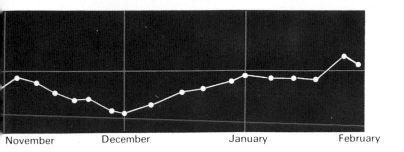

November   December   January   February

chances of finding a faint comet are much higher than for novae (M. P. Candy once found a comet when testing an eyepiece). A telescope with low magnification and wide field of view is required, and large binoculars are very useful.

Meteors. The apparatus needed here is simple – a good star chart or atlas, a watch (correct to the nearest minute), a ruler, and a blanket (to keep warm). Sporadic meteors can be seen on any night and meteor showers on certain occasions. When a meteor is observed, its brightness, any interesting features, the time, and its path (using the ruler held at arm's length) should be noted.

Aurorae. These very beautiful phenomena are best seen from polar latitudes, and of course near to the sunspot maximum. They can be studied either visually or photographically. Useful points to note are the forms displayed, such as rays, curtains and arcs (and in the latter case, the height of the arc above the horizon), as well as colours, which are often very delicate hues, with greens and pinks predominating, and the time and duration of the display.

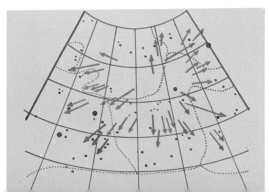

The paths of Perseid shower of meteors plotted on a meteor chart

153

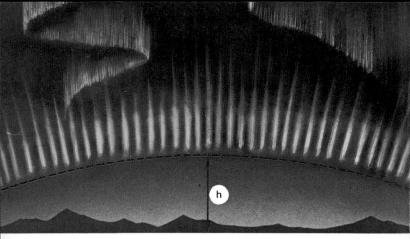

Auroral measurements. Often an arc-like band or 'rayed arc' is seen. The height *h* in degrees is required as well as colour and other details.

Radio astronomy. Even radio astronomy is open to the amateur in a small way. It is not at all hard to build suitable aerials to detect radio waves from the Sun and from Jupiter for example, although equipment such as amplifiers and pen recorders are needed to record the signals.

Astronomical societies

It stands to reason that observations made by a lone amateur are of little value except to himself. However, most countries have national amateur astronomical societies which correlate observations and publish the results, and it is well worth joining one, even if you have no intention of making serious

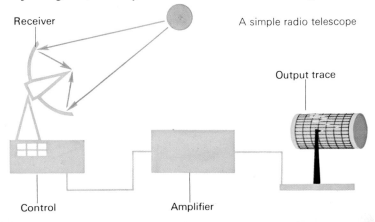

Receiver

A simple radio telescope

Output trace

Control

Amplifier

observations; they are a good means of meeting other amateurs and keeping up-to-date on new developments. Some societies, such as the British Astronomical Association, have telescopes which are available for loan to members. Many towns have their own astronomical societies, often with reasonably large telescopes.

The planetarium

There are planetaria in many large cities throughout the world; the first was set up at Jena in Germany by the Zeiss optical company. The interior of these buildings consists of a dome on which the stars, planets, Sun and Moon are projected and made to go through their motions while the lecturer describes what is going on. Eclipses can be arranged, as well as comets and other features, and a planetarium is well worth a visit should the opportunity arise.

The Zeiss planetarium projector

BOOKS TO READ

For general introductions to astronomy the following books are recommended and are usually available from bookshops and public libraries. These are only a very small selection from the large number of good books available on astronomy related topics.

The Planets, Heather Couper and Nigel Henbest, Pan Books, London, 1985.
The Stars, Heather Couper and Nigel Henbest, Pan Books, London, 1988.
Yearbook of Astronomy, edited by Patrick Moore, Sidgwick & Jackson, London, yearly.
Cosmos, Carl Sagan, Macdonald, London, 1981.
Secrets of the Sky, Ian Ridpath, Hamlyn, London, 985.

ASTRONOMICAL SOCIETIES

The Junior Astronomical Society, 36 Fairway, Keyworth, Nottingham, NG12 5DU, England.
The British Astronomical Society, Burlington House, Piccadilly, London, W1V 9AG, England.
The Amateur Astronomy Centre, Clough Bank, Bacup Road, Todmorden, Lancashire, OL14 7HW, England.
The Astronomical Society of the Pacific, 1290 24th Ave., San Francisco, California 94122, USA.

PERIODICALS

In addition to the publications of the astronomical societies, you will gain a lot from reading such periodicals as:
Astronomy Now, Intra Press, London, monthly, through newsagents.
Astronomy, Astromedia Corp., American, monthly, through newsagents.

INDEX

Page numbers in bold type refer to illustrations

Absolute magnitude 79
Achernar 129
Achromatic lens 21
Adams, John Couch 19, 65
Aerolites 75
Albireo 137
Alcock, George 87, 137
Alcock's First Comet **152**
Alcor **82**, 83, 123
Aldebaran 12, 78, 93, 122, 129
Algol 85, 129
Almagest, The 11
Alphard 130
Altair 137
Andromeda **100**, 126
Andromeda Galaxy *see* Great Nebula in Andromeda
Antares 80, 134
Apollo spacecraft 52, 54, 116, **116**
Aquarius 126, 142
Aquila 137
Ara 134
Arcetri 42
Arcturus 121, 134
Arecibo 111
Argo Navis 130
Aries 126, 129, 142
Aristarchus 11
Aristarchus crater **54**
Aristotle 10, 12
Arizona Meteor Crater 74, **74**
Armillary sphere **12**
Asteroids 32 *see also* Minor planets
Astrolabe **12**
Astrology 8, 12, 143
Atmosphere 31, 49, **49**, 51, 112
Auriga 121, 122, 129
Aurora 47, **50**, 51, 153, 154

Barlow lens 148
Barycentre **50**
Beehive 130
Bessel, F. W. 19
Betelgeuse 80, 85, 89, 122, 129
Biela's Comet 73
Big Dipper 121, *see also* Ursa Major
Binoculars 145, **145**

Bode's law 68
Bondi, H. 107
Boötes 121, 134
Bridal Veil Nebula **18**
British Astronomical Assoc. 155
Brooke's Comet **70**
Bull *see* Taurus

Camelopardalis 123
Cancer 130, 142
Canes Venatici 93, 133
Canis Major 122, 130
Canis Minor 122, 130
Canopus 140
Capella 89
Capricornus 137, 142
Captured rotation 52
Carina 130, 140
Cassini Division 63
Cassiopeia 121, 123, 126, 129
Castor 122, 130
Celestial equator 76
Celestial police 68–69
Celestial sphere 76
Centaurus 133
Cepheus 121, 123
  δ Cephei 85
  Cepheid variables 85–86, 91, 100, 103
Ceres 69
Cetus 85, 122, 126
Charioteer *see* Auriga
Chromatic aberration **20**, 21, 147
Chromosphere 43, 44
Circumpolar stars 118
Coal Sack **94**, 95
Colour-magnitude diagram 91
Columba 129
Coma 72
Coma Berenices 133
Comets 6, 32, 70–73, **73** 152
Conjunction 32, **144**
Constellations 118–140
  maps **120–137**
Copernicus 13, 14
Copernicus crater **52**
Corona 41, 44, 47
Corona Borealis 134
Coronagraph 43
Corvus 133
Crab *see* Cancer
Crab Nebula **87**, 129
Crepe Ring 63

Crow *see* Corvus
Crux' 140
Cusp caps 149
Cygnus 121, 137
  61 Cygni 19, 137

De Cheseaux' Comet 71
Declination 76, **119**
Deimos 59, **59**
Delphinus 87, 137
Deneb 137
De Revolutionibus Orbium Coelestium 13
Dichotomy 149
Doppler effect 81, 82
Dorado 140
Draco 9, 123
Dragon *see* Draco

Early Bird 112–113
Earth 4, 5, 6, 10, 11, 13, 14, 17, 46, 48–52, 67, 68, 112, 118
  age of 103
  structure of **48**, 49
Eclipse 7, 10, 34, **34**, 35, 43, 44
Ecliptic 142
Egyptians 8
Egyptian calendar 9
Emission lines 43
Eratosthenes 10
Eridanus 129
Evolutionary theory 106–107, 108

Faculae **40**, 41
First Point in Aries 126, 142
Fishes *see* Pisces
Fomalhaut 126
Fraunhofer lines **26**, 27, 42, 43, 45, 81, 83

Galaxies 98, **98**, 99, 100–109, **101**
Galileo 15–16, 21, 40, 61, 62
  telescope of **14**
Gemini 121, 122, 130, 142
Gold, Thomas 107
Gravitation 17
Great Bear 83, 120, 134, *see also* Ursa Major
Great Comet **70**
Greatest elongation 33

Great Looped Nebula 140
Great Nebula in Andromeda **18**, 100–101, 103, 126
Great Nebula in Orion **94**, 95, 129
Great Pyramid of Cheops 9, **9**
Great Red Spot 60, 61, **61**
Greek Astronomers 10
Grus 126

Halley's Comet 70–71, **71**, 72
Halley, Sir Edmond 70–71
Hare *see* Lepus
Heliocentric theory 14
Hercules 81, 93, 134
Herdsman *see* Boötes
Hermes 69
Herschel, W. 19, 64, 100
  telescope of **22**
Hertzsprung- Russell diagram **90**, 91
Hevelius 21
  telescope of **17**
Hipparchus 10, 11
Horse's Head Nebula **96**
Hoyle, F. 107
Hubble 100, 105
Hubble's constant 105, 109
  diagram **104**
Huggins 19
Huyghens 62, 63
Hyades 129
Hydra 122, 130, 133
Hydrus 140

Interstellar dust 96–97
Interstellar gas 96–97
Interstellar reddening, graph of 96
Ionosphere 51

Jansky, Karl 29
Janus 63
Jeans, Sir James 36
Jodrell Bank 29, **29**
Juno 69
Jupiter 11, 16, 32, 60–61, **60**, 67, 68, 73, 144, 150
  belts **151**

Kepler, J. 14, 15, 16, 68
Kepler's laws 15, **15**, 35
Kirchhoff, Gustaff 27
Koppernigk, N. *see* Copernicus
Kuiper's star 81

Lagoon Nebula 137
Laplace 36
Lemaître, George 106
Leo 121, 122, 123, 142
Leo Minor 133
Lepus 129
Le verrier 19, 65
Libra 134, 142
Lick Observatory 24
Light 26–28
Light year 79
Lippershey 15, 21
Lowell Observatory 105
Lowell, Percival 66
Lunar Orbiter 52, 113
Lunik 52, 113
Lupus 134
Lynx 130
Lyot monochromatic filter 43
Lyra 83, 1-21, 137
  RR Lyrae 85, 86, 93

Magellanic Clouds 101, 111
  Large 101, 140, 141, **141**
  Small 101, 140, 141
Main sequence 91
Mare Imbrium **52**
Mariner spacecraft 59, 114, **115**
Mars 11, 32, 58–59, 67, 68, 114, 116, 117, 150, **150**
  canals 59, 117
  surface **58**
Mercury 11, 32, 35, 55, **55**, 67, 145
  transit of **55**
Messier's Catalogue 95
Meteorites 51, 74–75, **75**
Meteors 32, 51, **72**, 73–75, 145, 153
  chart **153**
  Leonids 153
  showers 73, 153
Meudon Observatory 24
Milky Way 4, 5, 16, 76, **93**, 95, 123, 129, 137, 141, 145
Minor planets 32, 68–69, **69**
Mira **84**, 85
Mizar **82**, 83, 121, 123
Monoceros 130
Moon 6, 8, 10, 11, 16, 31, 34, 35, 51, 52–54, **52**, 56, 113, 114, 142, 147, 149
  landing 116
  lunar domes 149
  manned base 117
  phases 33, **33**
  surface features 53, **114**

Nebulae 95, 100
  dark 95, 96
  emission 95
  reflection 95
Nebular hypothesis 36, 37
Neptune 19, 32, 65, **65**, 67, 71
Newton, Isaac 16–17, 21, 22
Northern Cross 137
Northern Crown *see* Corona Borealis
Northern lights 51
Novae 86, 87, 91, 145, 152
  in Delphinus **86**
  in Hercules **86**
Nubecula Major 141, *see also* Magellanic Cloud, Large
Nubecula Minor 141, *see also* Magellanic Cloud, Small

Observatories, solar 42
Octans 140
Omega Nebula 137
Ophiuchus 134
Opposition 33
Optical doubles 82
Orion 85, 95, 120, 122, 129,

Pallas 69
Palomar Mountain 19, 24, **25**, 79, 103. 109
Parallax, method of **18**, 19, 86
Paris Observatory 17
Parsec 79
Pavo 140
Pegasus 126
Penumbra 40
Period-luminosity law 86
Perseus 85, 129
Phobos 59, **59**
Phoenix 126
Photography 148, **149**
Photosphere 40, 43
Pic du Midi Observatory **31**
Pickering 89
Pisces 76, 126, 142
Piscis Australis 126
Planetarium 155, **155**
Planets 7, 48–67, *see also individual planets*
Pleiades **92**, **94**, 95, 122, 129
Plough 120, 123, *see also* Ursa Major
Pluto 19, 32, 66, **66**, 67
Polaris 9, 121, 123
Pole Star 9
Pollux 122, 130

Praesepe 130
Precession 9, **9**, 11
Procyon 122, 130
Prominences 41
Proxima Centauri 133
Ptolemy 11, 12, 118
   universe of **11**
Pulsars 81
Pulsating universe **107**
Puppis 130

Quadrant **14**
Quasars 109, **109**

Radial motion 81
Radiation belts 47
Radiation, solar 38, 46
Radio astronomy 29–30, 154
Radio galaxies 108
Radio interferometer 30, **30**
Radio telescopes 97, 111
Ram *see* Aries
Red shift 105
Regulus 133
Rigel 122, 129
Right ascension 76
Ring Nebula 137
Royal Observatory,
   Greenwich **16**, 17
   Herstmonceux 24

Sagitta 137
Saggittarius 137, 142
Satellites 112–113, 114
Saturn 11, 19, 32, 62–63,
   67, 151
   rings of 62–63, **62**, **151**
Schroter effect 149
Scorpius 134, 142
Serpens Caput 134
Seven Sisters 93
Shooting stars 51, *see also*
   Meteors
Siderites 75
Sirius 8, 79, **82**, 89, 122, 130
Slipher 105
Southern Cross 95, *see also*
   Crux
Southern lights 51
Spacecraft 112–117
Spectrogram **26**, 28
Spectrograph 28
Spectroheliogram **44**
Spectroheliograph 42, 43, 44
Spectroscope 28, **28**
Spectrum 27
   flash 44
   stellar 88–89, **89**
Spica 121, 133

Sputnik 112, **112**
Stars, binary 82
   brightness 78–79
   catastrophic variables 86
   Cepheid variables 85–86,
   91, 100, 103
   classes 89
   clusters 93
   colours 80
   distance 79
   double 82–83
   eclipsing binary 85
   evolution 91, **91**
   flare 86
   giants 91
   irregular variables 85, **152**
   long-period variables 85
   magnitude scale 78–79
   masses 80
   optical doubles 82
   population I 99, 100
   population II 99, 100
   red giants 91
   RR Lyrae 85, 86, 93
   sizes 80
   structure 76
   supergiants 91
   variables 85–87, 151
   W Virginis 86
   white dwarfs 80
Steady-state theory **106**,
   107
Straight Wall **54**
Stromlo, Mount, Observatory
   111
Summer Triangle 137
Sun 4, 6, 8, 11, 13, 14–16,
   19, 27, 36–37, 38–47, **43**,
   51, 72, 79, 80, 81, 88, 89,
   91, 98, 99, 148, **148**
   eclipse of 34
   formation 47
   prominences 41, **41**
   radiation **46**
   solar constant 46
   solar observatories 42
   solar system 32–37
   solar wind 47
   structure 43
Sundial 7, **7**
Sunspots 39, 40–41, **40**, 46,
   147, 148, **148**
Supernovae 86, 87, 91
Superior planets 33

Taurus 12, **92**, 93, 122, 129,
   142
Tektites 75
Telescopes 15–17, 110, 111
   aerial 21
   distribution **111**

Herschel's telescope **22**
   radio 29, 30
   reflecting 16, **16**, 22–23,
   22, 147
   reflecting, home-made **146**
   reflecting, Schmidt **22**, 23
   refracting **20**, 21, **146**, 147
   solar **42**
Telescope mountings 24, 147
   altazimuth 24, **145**
   equatorial 24, **25**
Telstar 112
Thales 10
Thuban 9
Tides **50**, 51
Tiros **112**
Titan 63
Transit 35
Transverse motion 81
Trapezium **83**, 129
Triangulum 129
Triangulum Australe 140
Tucana 140
Twins *see* Gemini
Tycho Brahe 14
   observatory **13**

Umbra 40
Universe, theories of 106–
   108
Uranus 19, 32, 64, **64**, 67,
   145
Ursa Major **82**, 120, 123
Ursa Minor 121, 123

Van Allen radiation belts 112
Vega 137
Vela 130
Venus 8, 11, 32, 35, 56–57,
   67, 71, 79, 114, 116, 149
   ashen light 57
   phases 56
Vernal equinox 76, 142
Vesta 69
Viking spacecraft **150**
Virgo 121, 133, 142
Volans 140

Whale *see* Cetus
Wilson, Mount 19, 24, 42,
   100
Winged Horse *see* Pegasus
Woolfson, M. M. 37

Yerkes telescope 21, 24, **25**

Zodiac 142–143, **142**–**143**